Carlaile Largura do Vale
André Jun Miki
Débora Zerbinato

Production Automation

Carlaile Largura do Vale
André Jun Miki
Débora Zerbinato

Production Automation

Production Engineering Research and Reflections

ScienciaScripts

This book is a translation from the original published under ISBN 978-613-9-70402-6.

Publisher:
Sciencia Scripts
is a trademark of
Dodo Books Indian Ocean Ltd. and OmniScriptum S.R.L publishing group

120 High Road, East Finchley, London, N2 9ED, United Kingdom
Str. Armeneasca 28/1, office 1, Chisinau MD-2012, Republic of Moldova, Europe
Printed at: see last page
ISBN: 978-620-8-20399-3

Professor M.e. Carlaile Largura do Vale, Computer Scientist in the Production Engineering course at the Federal University of Rondônia (UNIR). carlaile@unir.br

Prof Dr André Jun Miki, PhD in Production Engineering: Energy, Environment and Clean Development Mechanism from the Methodist University of Piracicaba (2015), with a thesis entitled: Generation and use of photovoltaic energy in an isolated community in the state of Amazonas, using a life cycle approach. Master's degree in Environmental Sciences and Sustainability in Amazonia from the Federal University of Amazonas (2000). Graduated in Teacher Training in Specialised Subjects from the Federal Technological University of Paraná - Degree in Electrotechnics (1992). Adjunct Professor I at the Amazonas State University (2001-2010). Professor at the Federal University of Rondônia - Production Engineering - Cacoal - RO (2017). Collaborating Professor of the Professional Master's Programme in Public Administration - (PROFIAP) UNIR (2017). He has experience in Education and Computer Science. He also has experience in Electrical Engineering, with an emphasis on Electrical Systems and Public Policy.

Débora Zerbinato, a graduate student in Production Engineering at the Federal University of Rondônia.

SUMMARY

CHAPTER 1

TECHNOLOGICAL IMPACT OF AUTOMATION IN A CERAMICS
ORGANISATION
Carlaile Largura do Vale
Guilherme Marques de Oliveira

SUMMARY

The general objective of this research was to observe the technological impacts on a ceramics organisation, and the specific objectives were to observe its production process and its perception of the changes arising from its automation in terms of technology. The methodology used was a bibliographical survey on automation and then an interview with the person in charge of production to gather information. In accordance with ethical standards, the name of the organisation will not be disclosed. Based on the data collected, time optimisation in the firing process was observed as an advantage of this technology.

Keywords: Automation; Ceramic processes; Optimisation.

1 INTRODUCTION

Automation is increasingly present in people's daily lives and is also being used in production processes to save time and resources for the organisation. The general objective of this research is to observe the technological impacts on a ceramics organisation, and the specific objectives are to observe its production process and its perception of automation

This study seeks to observe the impact that technology, in this specific case automation, has had on the organisation. In terms of methodology, the research is descriptive, seeking to provide a new view of an existing reality. In the first stage, data was collected on the subject in the form of a bibliographical review using books and scientific articles. The second stage involved a visit to the company in early 2016 for an interview to obtain data on automation and its production process with an employee in charge of production.

The theoretical framework will be presented, followed by the results and discussions, ending with the final considerations.

2 THEORETICAL FRAMEWORK

Topics relating to automation itself will be listed, as well as its history and related areas such as production systems, software, Programmable Logic Control (PLC) and the man/machine interface.

2.1 History of Automation

Since the dawn of time, humanity has been developing its creativity to develop processes and tools to help its activities. Automation gained momentum with the industrial revolution and its technologies and ended up maximising human productive capacity, thus generating better integration between man and machinery and/or their tools for gradual gains in the production of goods and/or services, as well as reducing production costs.

According to Quintela (1998), the Industrial Revolution gave rise to a common "factory system" for the United States, Great Britain and other countries. In the middle of that century, however, the manufacturing system in the United States took on a distinctive nature that transformed the growth of the United States into economic power. Between 1875 and 1889, shortly after this system began to dominate the American economy, per capita production of goods grew at an annual rate of 2%, compared to 0.03% in the first three quarters of the century, and the share of industry in the production of goods grew from 17% to 53% between 1839 and the end of the century.

According to Silveira (2003), relays were created and these electromechanical devices became factory control and automation equipment. The growth of industry and the use of steel instead of iron boosted industry and, together with other factors, gave it the title of the second Industrial Revolution. At the beginning of the 20th century, the concept of industry was already widespread, although the factory environments of the time did not have very elaborate automation processes. In 1968, with the aim of replacing relays with a new electronic device, the North American company BedFord Association was responsible for designing and installing the new automation system. The MODICON (Modular Digital Controller) was the first Programmable Logic Controller to replace the entire automation system based on electromechanical relays, making the system much more flexible, economical and efficient.

2.2 Automation

According to Rosário (2009), industrial automation is the use of any mechanical or electro-electronic device to control machines and processes. With regard to electro-electronic devices, computers or other logical devices (such as Programmable Logic Controllers/Numerical Commands) can be used to replace the execution of tasks carried out by people or impractical for human labour. This represented a step beyond mechanisation, in which human employees are provided with machines to help them perform their tasks.

For Silveira (1998), automation is a set of techniques that favour the construction of active systems capable of acting with optimum efficiency through the use of information collected from the environment in which they operate. According to Rosário (2009), the concept of automation is often confused with that of automation. Automation has its concept linked to the execution of automated, mechanical and repetitive movements, being synonymous with mechanisation, a mechanism of blind implication and without correction. On the other hand, automation has the concept of a set of techniques that build systems capable of acting with optimum efficiency through the use of information collected from the environment in which it operates. Using this information, the system calculates the most appropriate corrective action. This gives the system a more balanced input/output relationship, thus correcting any values that are outside the desired values.

Automation makes extensive use of the same concepts as automation. However, according to Silveira (1998), the level of flexibility imputed to the system is much higher because it is inextricably linked to the concept of software. This resource provides the system that uses automation with the possibility of radically altering the entire automated behaviour in order to produce a different range of results.

2.2.1 Automation in the organisation

According to Rattner (1987 apud Silva, 2005), automation offers enormous advantages when compared to conventional electromechanical equipment, two of which can be mentioned: more efficient control and organisation of assembly lines; inventory-centred planning and control; computer-assisted design and testing; performance and equipment maintenance control; as well as an increase in the flexibility and reliability of equipment and machinery.

2.3 Production systems

There have been major advances in recent decades in production systems, which are divided into three classes: rigid, flexible and programmable. Rosário (2005) conceptualises rigid automation as being used for large volumes of production, making the production line fixed and turning its entire design towards a specific type of product. Flexible automation is geared towards medium-volume production, where automation is combined with flexibility, enabling industries to manufacture several products simultaneously using the same production system thanks to automation. Programmable production is differentiated from flexible production by the fact that it is produced in small batches, requiring the equipment to be reprogrammed to manufacture a new batch.

2.4 Software

This topic will cover the definitions of software, its nature in general, its emergence and the technological development it has brought about, as well as its field of application. Following the ideas of Pressman (2011), the term software refers to a sequence of printed or virtual descriptive logical instructions, a support for equipment of any size or architecture that allows data to be identified and manipulated and finally executed.

2.4.1 Programmable Logic Controller

This subject will cover the concept of Programmable Logic Control (PLC), its history, as well as its basic architecture and field of application. The Programmable Logic Controller (PLC) acts as a previously programmed digital logic system that informs and coordinates a piece of equipment or a group of equipment via input and output modules, as it would previously have required local control.

According to Ribeiro (2001), the PLC is an electronic computer that performs control functions of various levels and types of complexity in an operator-friendly way, and can even be programmed, controlled and operated by a lay person.

According to the Brazilian Association of Technical Standards (ABNT apud Parede, 2011), a PLC is "digital electronic equipment with hardware and software compatible with industrial applications". For the National Electrical Manufacturers Association (NEMA apud Parede, 2011), it is a: "digital electronic device that uses a programmable memory for the internal storage of instructions for specific implementations, such as logic, sequencing, timing, counting and arithmetic, to control through input and output modules various types of machines and processes".

2.4.2 Human-machine interface

Since the dawn of time, man has been limited in some activities, be they physical or mental, so he can't or can't perform a task that exceeds these limits in the way he wants. To make up for this deficit in human capacity, man created the machine, but what is a machine?

The word "machine" has a close link with the term "technology" because, being commonly related to Greek origin, technology according to Rodrigues (2001) means the reason for knowing how to do things. According to Andrade (2013), a machine is the set of mechanical and electrical devices responsible for operating the system. Therefore, the machine is the manifestation of man's reason and will in the realisation of a task or product.

According to Parede (2011), the human-machine interface (HMI) serves as a direct communication device with the PLC and is used to visualise process data, performing monitoring and control functions for machines, installations and industrial processes. In terms of functions, Ribeiro (2001) adds that these include access to read/write register data, programming and diagnostics. It is possible to interact with the PLC's internal registers and mesh tables, giving the system operator greater real-time tuning of the variables and better control of the system.

According to Costa (2011), the vast majority of automation systems have some kind of interface with the operator, in order to display process data and make changes to parameters. HMIs therefore play an important role in the operation and fault detection of a machine or process. This type of equipment has such a wide range of applications that it can be used in building and residential systems, as well as in commercial areas such as ATMs.

3 . RESULTS AND DISCUSSIONS

The research company was founded in the 1980s in the city of Cacoal, in the state of Rondônia, and has 110 employees. The company operates in the ceramics industry and sells the following products: double Portuguese and Roman roof tiles, ridge tiles and paulistinha roof tiles, 6-hole sealing blocks (bricks) measuring 90 mm x 140 mm x 240 mm, 6-hole blocks (bricks) 110.50 mm x 140 mm x 240 mm, with their respective % blocks, H7-type ceiling slabs, hollow elements and exposed bricks.

3.1 Production process

The production process of the company studied basically consists of the flowchart below.

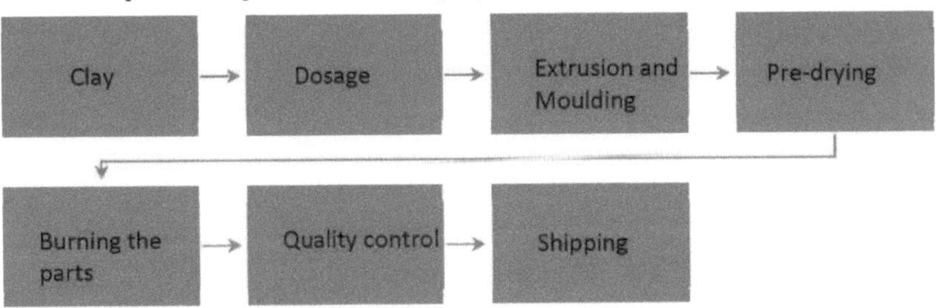

FIGURE 1. CERAMIC PRODUCTION PROCESS
SOURCE: PREPARED BY THE AUTHOR.

The clay has a resting time of approximately one year, where it interacts with the environment to obtain the characteristics needed for the next stages, including humidity. The next stage is dosing, where the composition of the mixture that will develop the mass for production is determined according to each product, be it bricks or tiles, thus varying the additives and binding materials for each specific product. The next stage is extrusion and moulding, where the mass comes out of the extruder through a dripper where vegetable oil is added to the blocks for the following processes, being pre-formed into blocks of standard dimensions according to the product's technical standards, in this process the air is eliminated making the mass more compact and homogeneous. This mass contains around 22% humidity and under these conditions it goes through forming, in which this piece of clay will obtain its final shape (tile, brick, etc.), by means of a cold press.

In the pre-drying stage, the piece will be exposed on a mobile structure with a ceramic base for around 36 hours, in order to lose more moisture and avoid possible future damage to the ceramic piece. The product goes through a further drying process, which is carried out in an oven, where the pieces enter with a humidity of 12% and leave the oven with a humidity of between 1.5 and 2%. This stage lasts 34 hours, at a temperature of 70 to 65°C. Before the next stage, the product undergoes quality control to check for defects such as missing pins, warping and cracks.

In the firing stage, the product inside the continuous kiln goes through a pre-firing process to lose more of the moisture present in the piece. It starts at an average temperature of 85 C, gradually increasing until it reaches a maximum temperature of between 750 and 800° C. The piece goes through a further quality control stage where its compressive strength, moisture capacity and bell sound will be measured. Finally, it is separated into batches and dispatched to consumers.

3.2 Automation in the Firing Process

It was observed that automation played an important role in the organisation, generating a reduction in the time required for firing, a crucial factor in increasing production at the company studied. Below is a table showing this variation in time. The company has been in business for just over thirty years and has used other types of kilns without automation, gradually switching to more efficient kilns as its history unfolded. The company needed one year to fully implement the automated continuous furnace.

TABLE 1: CORRELATION OF OVEN USE

Oven type	Time required for burning	Moisture in the product
Vault	8 days	10%
Hoffmann	6 days	5%
Continuous (tunnel)	39 hours	1%

SOURCE: PREPARED BY THE AUTHOR

It can then be seen that there was a significant reduction in the amount of time needed for the firing process, which went from 8 days to 39 hours, thus generating a greater quantity of products to be manufactured, as well as an increase in their quality due to the reduction in humidity found in the ceramic pieces.

4 FINAL CONSIDERATIONS

As for automation, the company believes that this technology is important for the organisation, but its only major negative point is the great distance from the after-sales representatives, which sometimes leads to setbacks due to abrupt stoppages. The company also has an automated cutting machine and has sometimes had unwanted downtime, which has delayed other processes until after-sales support has visited.

It can be seen that automation is increasingly present today, thus generating the need for organisations to use such technologies in order to optimise their processes and thus obtain competitive gains. This study only looked at the technological impacts of this automation, thus observing an increase in productivity generated by the implementation of automation in this process.

REFERENCES

ANDRADE, Willian Jeferson. **Industrial application of Bosch Rexroth programmable logic controllers, human machine interfaces and industrial computers.** Curitiba: UTFPR, 2013. Available at:

<http://repositorio.roca.utfpr.edu.br/jspui/bitstream/1/2409/1/CT_CEAUT_IV_2014_1   2.pdf>. Accessed on: 10 December 2015.

COSTA, Luiz Augusto A. **Specifying Industrial Automation Systems.** São Paulo: Biblioteca 24 horas, 2011. 1 Edition. 205p.

PAREDE, Ismael Moura. GOMES, Luiz Eduardo Lemes. HORTA, Edson. et al. **Electronics: industrial automation.** São Paulo: Fundação Padre Anchieta, 2011.

PRESSMAN, Roger S. **Software Engineering.** McGraw Hill Brazil, 2011.

QUINTELLA, Heitor M. **Production Automation and Organisational Change: Analysis Models and the Case of Brazil.** 1998. Available at: <http://www.abepro.org.br/biblioteca/enegep1998_art469.pdf>. Accessed on: 28 Nov. 2015.

RIBEIRO, Marco Antônio. **Industrial Automation.** 4 ed. Salvador: Tek, 2001.

ROSÁRIO, João Mauricio. **Industrial automation.** São Paulo: Baraúna, 2009.

ROSÁRIO, João Maurício. **Principles of mechatronics.** São Paulo: Prentice Hall, 2005.

SILEVIRA, Leonardo. LIMA, Weldson Q. **A brief conceptual history of Industrial Automation and Networks for Industrial Automation.** 2003. Available at: < http://www.dca.ufrn.br/~affonso/FTP/DCA447/trabalho1/trabalho1_13.pdf>. Accessed on: 28 Nov. 2015.

SILVA, Frederico Heitor Jesus. **Work organisation and the impact of automation in the automobile industry.** Ouro Preto: UFOP, 2005. Available

at:<http://www.em.ufop.br/cecau/monografias/2005/FREDERICO%20HEITOR%20
JESUS%20SILVA.pdf>. Accessed on: 24 Nov. 2015.

SILVEIRA, Paulo Rogério da. SANTOS, Winderson E. dos. **Automation and Discrete Control**. 9. ed. São Paulo: Érica, 1998.

CHAPTER 2

PRODUCTION AUTOMATION IN A CLOTHING INDUSTRY IN PIMENTA BUENO

André Jun Miki
Carlaile Largura do Vale
Jonathan Alves Santos
Jorge Luis Barbosa Habitzreuter

SUMMARY

In recent years, countless social, political, economic and technological changes have taken place, making significant modifications to the productive sectors necessary. The textile industry was established in Brazil in the mid-19th century, and this sector has traditionally helped to drive the development of industrialisation in other countries. The aim is to analyse the impacts of implementing and maintaining automated processes in a textile industry, identifying the benefits and difficulties of the automation process in the organisation and analysing the maintenance process in the use of equipment. Competition with imported and domestic products has led companies to acquire automated machinery and systems in order to improve the quality of their clothes, increase productivity and reduce costs. The state of Rondônia shows its potential when it comes to clothing and the local market is currently competing with industries in other regions of the country, such as Goiânia, Paraná and cities in the Northeast. This research is classified as descriptive, with a qualitative approach and the method used was deductive. The research used bibliographies obtained from books, articles and dissertations in the area under study, which led to a semi-structured interview with the company's production manager and a questionnaire applied to twenty of the organisation's employees. The data collected and analysed shows that the process of automating the company in question took a long time and involved a high level of investment. One of the biggest obstacles to achieving the desired objectives is the resistance of certain employees to adapting to new technologies.

Keywords: Automation, Clothing, Textile Industry.

1 INTRODUCTION

Man has always sought simpler, faster and more precise ways of doing his work. This can be seen in the development and creation of tools in the stone age, passing through various other inventions and culminating in the industrial revolution, a major milestone in which machines came to replace human labour once and for all. With it came mass production, the production line and many other concepts that are part of our daily lives.

The textile sector, including clothing and apparel, is of great importance to the international, national and regional economy and is a major generator of jobs (ARAÚJO; PEREIRA, 2006).

11

Traditionally, the textile sector has helped in the industrialisation process of many countries. Sachs (2005) shows that investing in the clothing sector is now a way in which nations plunged into absolute poverty can "step" onto the first rung of the development ladder, citing the case of Bangladesh.

The textile sector was the driving force behind the English industrial revolution of the 18th and 19th centuries. In Brazil it was no different, with the textile industry having been of great importance since before the 1950s (CAMPOS, 2005).

Throughout the 2000s, Brazil lost competitiveness and market share in the textiles and clothing sector. Despite strong growth in world consumption of textiles and clothing, the country's share of world trade declined from 0.7% in 1997 to 0.3% in 2007. In addition, global competition has intensified with the exponential growth of Asian products in international trade, especially China. In this context, it has become essential for the survival of companies in the textile and clothing chain to develop differentiated competitive strategies, based on the use of technological innovation as a relevant instrument for entering the world market (COSTA; ROCHA, 2009).

The textile sector is made up of around 5,000 companies, of which only 11 per cent are considered large and 21 per cent small and medium-sized. Micro-enterprises, which account for 68 per cent of the total, represent the vast majority of the sector (BORGES et al., 2005).

The development of robotics, automation and the phenomenon of globalisation, which expanded capitalist competition from the local to the global level, were the foundations for the transformations that took place in the textile industry. With globalisation, the Brazilian consumer market now has access to fabrics from other countries, of superior quality and at more affordable prices. This phenomenon required the replacement of machinery and administrative modernisation (SOUZA and ANDRIOLA, 1999).

2 THEORETICAL FRAMEWORK

This topic will present some characteristics about the history of automation, automation, production systems, the structure of the PLC (Logic Programming Controller) and the advantages of using industrial automation in the textile sector. With the aim of developing ideas based on bibliographical references guiding the research, presenting a foundation of literature already published on the same subject.

2.1 History of Automation

According to Abimaq (2006), since prehistoric times, human beings have somehow processed stones, then metals, then more and more elaborate pieces until they reached the construction of simple and efficient machines, but with manual propulsion. That's why they weren't yet considered machine tools, machines capable of extending intelligent human action without their own energy.

It's not so easy to pinpoint the emergence of Industrial Automation, but it is known that for there to be industrial automation there must first be industry, and also self-controllable automatic processes. In this way, the beginnings of Industrial Automation can be traced back to the 18th century, with the British creation of the steam engine, increasing the production of manufactured goods, and these were the decades of the Industrial Revolution. In the following century, industry rose and took shape, new energy sources and the replacement of iron with steel fuelled the development of industries in Europe and the USA. In this context, in the years that followed, mechanical devices called relays were created, which would soon take over factories. All these events, and others that followed, were called the Second Industrial Revolution (SILEVIRA and LIMA, 2003).

The concept of mechanisation can be defined as the use of machines to replace human or animal labour. The scientific discovery of the elastic force of water vapour, which triggered a process of industrial renewal, dates back to the 18th century and completely changed the structure of the countries where it took place. According to Scopel (1995), it was in England, around 1760, that the first machines powered by this energy were put into operation. The industries that pioneered the use of this new form of energy were the spinning mills, which immediately entered the era of mass production. With the advent of new inventions and their dissemination, new development processes emerged, while the hope of better wages for workers and their families was eliminated, as machines emerged with production capacities up to 120 times greater than those provided by the old methods. Muscle labour was finally replaced by machine labour (SCOPEL,1995).

2.2 Automation

In an industrial context, automation can be defined as the technology that deals with the use of mechanical, electro-electronic and computer systems in the operation and control of production (PAZOS, 2002).

Ribeiro (1999) says that automation is the exchange or substitution of functions performed by humans or animals for machines or mechanical processes. Automation is the use of machines or automatic processes that are controlled by some device, with minimal contact with the system controller. It is the control of automatic processes. Automatic is having its own operating characteristics, thus carrying out its action according to the determinations that have been laid down for that system.

Automation is a set of techniques designed to make tasks automatic, replacing the expenditure of human bioenergy, with muscular and mental effort, with computable electromechanical elements (SILEVIRA and LIMA, 2003). When automating a production process, it is necessary to use mechanical, electrical and electronic devices that perform functions equivalent to those of humans in supervision and control activities, such as collecting and analysing data and correcting course (GUTIERREZ and PAN, 2008).

2.3 Production Systems

A system can be defined, according to Chiavenato (1983) and Ballestero-Alvarez (1990), as a set of interacting and interdependent parts (or elements or organs), i.e. dynamically interrelated, which together form a unified whole, and which carry out an activity or function to achieve one or more objectives or purposes (system purpose).

A Production System can be defined as a "set of interrelated activities involved in the production of goods (in the case of industries) or services" (MOREIRA, 2000).

Moreira (2002) explains that a 'production system' is the set of interrelated activities and operations involved in the production of goods (in the case of industries) or services. The production system is still an abstract entity, but it is extremely useful for giving an idea of totality. Production systems, according to Kopak (2003), as well as providing support so that companies can achieve their strategic objectives, must be able to support the logistics decision maker to:

A. Planning the materials purchased, the organisation's future production capacity needs, the appropriate levels of stocks and raw materials, semi-finished and finished products, in the quantities required;
B. Scheduling production activities;
C. Providing correct information on the current status of production resources and orders, both purchasing and production, so that orders can be delivered to customers as quickly as possible.

Slack, Chambers and Johnston (2002) explain that any operation that produces goods or services, or a mixture of the two, does so through a process of transformation and, by transformation, they mean the use of resources to change the state or condition of something in order to produce outputs. In short, production involves a set of input resources used to transform something or to be transformed into outputs of goods or services.

A Flexible System is a grouping of semi-independent computer-controlled workstations linked by an automated transport (or handling) system. Its implementation is recommended when there is a high variety of parts to be produced, in low and medium production volumes (MARTINS,

2002).

Tubino (1997) discusses the classifications of production systems more broadly and identifies the criteria on which three of them are based:

1. By the degree of standardisation
   a. Systems that produce standardised products: goods or services that have a high degree of uniformity and are produced on a large scale;
   b. Systems that produce customised products: goods or services developed for a specific client.
2. By type of operation
   a. Continuous processes: involve the production of goods or services that cannot be identified individually;
   b. Discrete processes: involve the production of goods or services that cannot be isolated, in batches or units, and identified in relation to others. They can be subdivided into
      i. Mass repetitive processes: large-scale production of highly standardised products;
      ii. Repetitive batch processes: production in batches of an average volume of standardised goods or services;
      iii. Project-based processes: meeting a specific customer need, the product designed in close liaison with the customer has a set date for completion. Once completed, the production system turns to a new project.
3. By the nature of the product
   a. Manufacturing of goods: when the product manufactured is tangible;
   b. Service provider: when the product generated is intangible.

2.4 The Structure of the PLC (Logic Control Programming)

The Programmable Logic Controller (PLC) was born in the North American car industry, specifically in GM's hydraulics division in 1968. Under the leadership of engineer Richard Morley, a specification was prepared that reflected the feelings of many users of relay controls, not only in the automotive industry but in industry in general. This sentiment stemmed from the great difficulty of changing the process using relay controls. Each significant change to a car model required alterations that added, removed or modified certain steps in the process, and this required changing all the panels and field wiring. In addition, the complexity and large size of relay panels made maintenance difficult (KOPELVSKI, 2010).

According to Kopelvski (2010), the concept of a programmable controller is quite broad. A programmable controller is nothing more than a computer with a physical construction that meets the requirements for operation in industrial environments. In addition, it has specific software for automation and control and often has a real-time operating system with a fixed memory configuration. In recent years, however, controllers based on standardised buses, general-purpose operating systems and an open memory structure have become increasingly popular.

2.5 The Advantages of Using Industrial Automation in the Apparel Sector

Industrial automation is a way that many companies have found to improve the production process of their products. One of the advantages of using industrial automation is the fact that machines, combined with technological advances and information technology, can do a man's job better and faster (PORTAL EDUCAÇÃO, 2015).

In Brazil, the textiles sector initially played a major economic role, followed by strong industrial development. Souza (2006) states that due to the low technological demand in relation to other industrial segments, the implementation of textile activities in Brazil took place before other industrial activities. Other factors that contributed to the implementation of these industries were that Brazil already had a cotton crop, an abundant labour force and a growing consumer market. However, the consolidation of the Brazilian textile industry took place during the First World War, when Brazil adopted an import substitution strategy. There was a period of relative industrial stagnation between the 1929 crisis and the Second World War, and then a period of technological renewal from the 1950s onwards. Among the innovations was the use of chemical products.

One of the basic characteristics of modern manufacturing is the search for production flexibility. According to Tavares (1990), the garment production process can be schematically divided into three stages: the conception and development of the product; pre-moulding, which involves graduating the base mould into the different sizes to be manufactured; the pattern through sewing and other thermal or chemical processes, followed by ironing or pressing, inspection and packaging.

According to Tavares (1990), there are two types of technological change in the sector:

A. Automation, which consists of equipping machines with certain automatisms and devices that allow machine time and manual operating time to be overlapped. They reduce the need for permanent control of the process by the operator.

B. Computerisation, which consists of using a calculator/simulator for certain stages of the production process. The computer then began to be used to manage orders and stocks and to monitor the flow of production in real time.

3 RESULTS AND ANALYSIS

To collect the data, visits were made on 19 and 20 January 2016 to a clothing manufacturer located in the city of Pimenta Bueno. A semi-structured interview was conducted with the company's production manager and a questionnaire was administered to twenty of the organisation's employees.

3.1 History of the organisation

The company studied in this research is a clothing manufacturer located in the city of Pimenta Bueno. The first steps were taken informally by the matriarch of the family in 1986 as a seamstress. It was only in 1996 that the company was formalised. Today the organisation has 85 direct employees.

The industry has most of the appropriate technological equipment for its production. Its initial activities were located in a small factory, but it has now moved to a new location with a more favourable environment for production. According to the production manager, labour is in short supply in the sewing sector.

3.2 Interview with manager

A semi-structured interview with the production manager on 20 January 2016 revealed some relevant aspects of the company's automation process.

When asked if there had been an increase in productivity following the implementation of the automation system, the manager was clear and objective, saying that there had been numerous benefits in the production process, she measured this improvement scale at a level 5, i.e. a lot of improvement. She also stressed that there had been a considerable increase in the final quality of the parts, and that production losses had been minimised to a surprising degree.

When asked about the difficulties encountered in the process of automating production, the production manager said that in 2016 the company had been in existence for two decades. It began in the owners' home in a very manual and simple way, it was a family business that worked in a very artisanal way and had almost no automated equipment. Over time, demand for the products began to increase and the first industrial sewing machines were introduced, new seamstresses were hired and the organisation began to grow. Over time, they managed to rent premises that would serve as a base for the clothing industry. Over the course of these 20 years, the company has consolidated its position in the industry and is now considered one of the largest in the city of Pimenta Bueno. The automation process had a boom in 2014 when they moved into a spacious and modern building, new equipment was acquired and technologies implemented, including a highly efficient laser cutting machine, new modern sewing machines, a pattern creation room with digitising tables and a professional printer for printing patterns, among other improvements. This whole automation process was extremely costly, and many difficulties were encountered, including the resistance of the seamstresses to handling the new sewing machines, "we had to train and qualify all our employees so that they could use all this new equipment more appropriately". This whole process took a long time, but with the automation that took place in 2014, the company has seen an increase in sales and profits, and is already looking at new machinery to further modernise its production process.

For her, the benefits obtained have been very satisfactory and the automation has made the process faster, the cuts in the fabrics are made more precisely because the laser cutting machine has a high level of efficiency, avoiding losses at this stage of the process.

As for the level of difficulty encountered in installing the automated system, she argued that several difficulties had been encountered, such as the high cost of setting up this system, the long time taken and the lack of qualified labour,

3.3 Questionnaire with Operators

Twenty employees were interviewed for the study in question, eighteen of them seamstresses and two cutting machine operators. Below are the results found and the interview questionnaire in Appendix A.

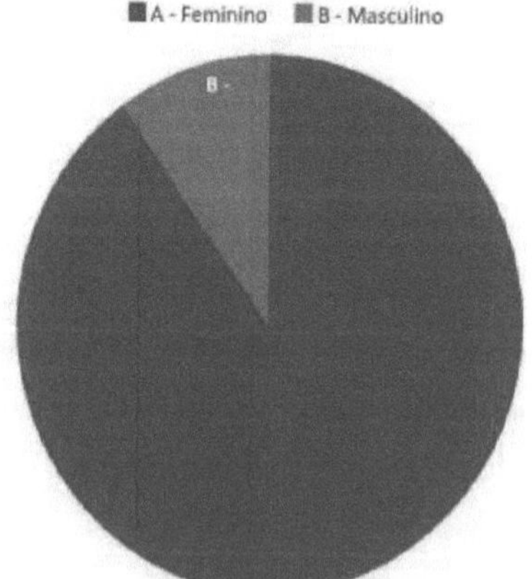

GRAPH 1 - SEX OF INTERVIEWEE
SOURCE: PREPARED BY THE AUTHORS, 2016.

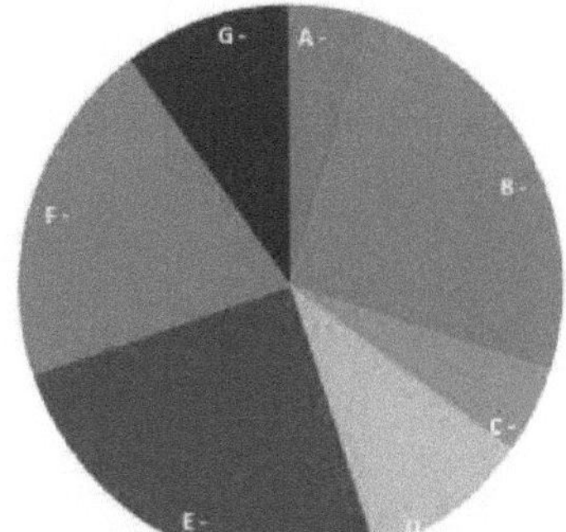

GRAPH 2 - AGE GROUP

SOURCE: PREPARED BY THE AUTHORS, 2016.

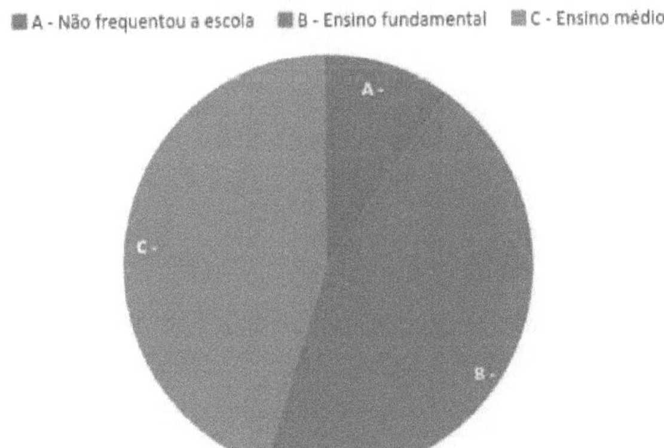

GRAPH 3 - LEVEL OF EDUCATION

SOURCE: PREPARED BY THE AUTHORS, 2016.

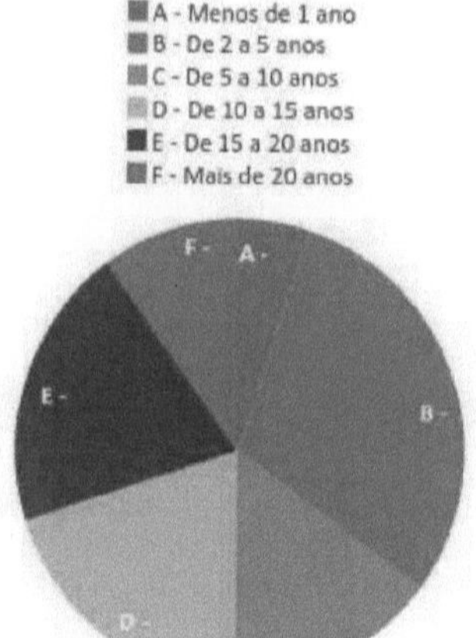

GRAPH 4 - LENGTH OF TIME IN THE ORGANISATION

SOURCE: (PREPARED BY THE AUTHORS, 2016).

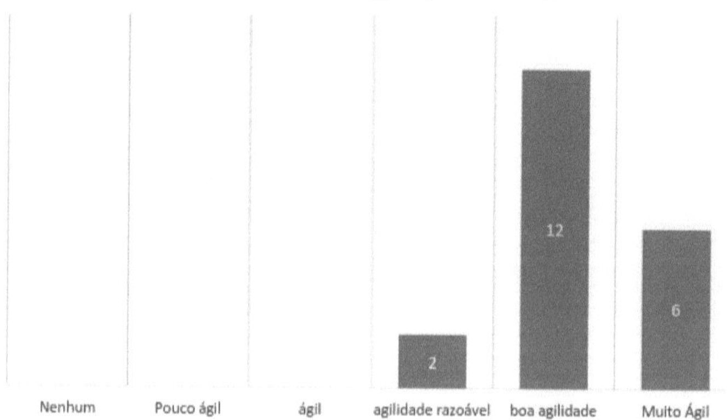

GRAPH 5 - PROCESS AGILITY

SOURCE: (PREPARED BY THE AUTHORS, 2016).

Graph 5 shows that employees believe the production process has become more agile since the automation process. According to the manager, there have been numerous benefits in the production process, she measured this improvement scale at a level 5, i.e. a lot of improvement.

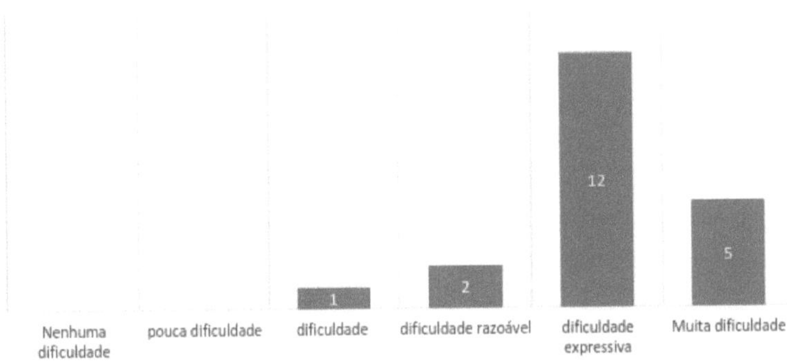

GRAPH 6 - **DIFFICULTY** ADAPTING

SOURCE: (PREPARED BY THE AUTHORS, 2016).

Graph 6 shows that employees found it difficult to adapt to the new equipment. For the manager, an example of the difficulty in adapting was the resistance of the seamstresses to handling the new sewing machines, which required training.

Dent and Golberg (1999) suggest that change agents, when proposing change, consider the existence of resistance and then include this circumstance in the planning to create mechanisms to deal with it.

When processes follow a pre-established routine, people need little effort to act and interact in organisations, but when change is proposed, it requires alterations in habits and attitudes, which generates discomfort (SCHWARTZ; ROCK, 2007).

4 FINAL CONSIDERATIONS

From the data collected and the analyses carried out, it can be concluded that the positive impacts following the implementation of the automated processes were an increase in the final quality of the garments, a reduction in losses resulting from large-scale production, a faster process and more precise cuts in the fabrics.

At the same time as the negative impacts, the implementation of the automation process was extremely costly, and after it was installed the seamstresses' resistance to handling the new sewing machines was another impact, with the need to train and qualify all our employees so that they could use all this new equipment more appropriately.

5 REFERENCES

ABIMAQ. **The History of Machines.** São Paulo, 2006.

ARAÚJO, C. A. L.; PEREIRA, C. F. **The clothing industry: impacts and opportunities in a post-ATC (Agreement on Textiles and Clothing) scenario.** XIII SIMPEP. Bauru, 2006.

BALLESTERO-ALVAREZ, M. E. **Organisation, systems and methods.** São Paulo: Mc graw Hill, 1990.

BORGES, C. R.; KALATZIS, A. E. G.; VELHO, P. R. **Promotion strategies for Brazilian foreign trade:** the case of the textile sector. Uniara Magazine, 2005/2006. Available   at: <http://www.uniara.com.br/legado/revistauniara/pdf/17/rev17completa_13.pdf> Accessed on: 12 Dec. 2015.

CAMPOS, L. H. R.; CAMPOS, M. J. C. **Competitiveness of the Brazilian Textile Sector:** a state-level approach. Fortaleza: BNB, 2005.

CHIAVENATO, I. **Introdução à Teoria Geral da Administração.** 3 ed. São Paulo:

COSTA, A. C. R; ROCHA, E. R. P. **Overview of the textile and clothing production chain and the issue of   innovation.**   Available   at: <http://www.bndes.gov.br/SiteBNDES/export/sites/default/bndes_pt/Galerias/Arquivo A/conhecimento/bnset/Set2905.pdf> Accessed on: 10 Dec. 2015.

DENT, E. B., GOLDBERG, S. G. **Challenging 'resistance to change'.** The Journal of Applied Behavioural Science, Thousand Oaks, v.35, n.1, p. 25-41, 1999.

GUTIERREZ, R. M. V.; PAN, S. S. K. **Electronic complex:** industrial control automation. Available at   : <http://www.bndes.gov.br/SiteBNDES/export/sites/default/bndes_pt/Galerias/Arquivo A/conhecimento/bnset/set2807.pdf> Accessed on: 15 January 2016.

KOPELVSKI, M. M. **Theory of PLC.** Federal Institute of Education, Science and Technology of São Paulo, 2010.

MARTINS, P. G., LAUGENI, F. P. **Administração da Produção.** 1ª ed. São Paulo: Saraiva, 2002.

MOREIRA, D. A. **Production and Operations Management.** 5 ed. São Paulo: Pioneira, 2000.

MOREIRA, D. A. **Production and operations management.** São Paulo: Pioneira

Thompson Learning, 2002.

PAZOS, F. **Systems automation & robotics.** Rio de Janeiro: Axcel Books, 2002.

EDUCATION PORTAL. **Learn about the advantages of industrial automation.**
Available at   :
<http://www.portaleducacao.com.br/informatica/artigos/53605/conheca-as-   advantages-and-
disadvantages-of-industrial-automation> Accessed on: 15 Dec. 2015.

SACHS, J. **Pedro Maia Soares, O fim da pobreza.** São Paulo: Companhia das letras, 2005.

SCHWARTZ, J.; ROCK, D. **The neuroscience of leadership.** HSM Management. Jan/Feb, 2007.

SCOPEL, L. M. M. **Automação industrial: uma abordagem técnica e econômica.** Educs, 1995.

SLACK, N.; CHAMBERS, S.; JOHNSTON, R. **Administração da Produção.** 2 ed. São Paulo: Atlas, 2002.

SOUZA, E. **Entrepreneurial Behaviour and Growth of Ceará Clothing Companies.** Fortaleza, 2006. 95 p. Monograph (Graduation) - Faculty of Economics, Administration, Actuarial Science and Accounting - Federal University of Ceará.

SOUZA, W. J.; ANDRIOLA, I. R. F. **O Homem e a Máquina:** Um Estudo das Imagens e Representações da Automação no Setor Industrial Têxtil. Federal University of Ceará, 1999.

TUBINO, D. F. **Manual de Planejamento e Controle da Produção.** São Paulo: Atlas, 1997.

CHAPTER 3

THE CHALLENGES OF ROBOTICS IN THE TEACHING AND LEARNING SYSTEM

Acsa Pires de Souza
Carlaile Largura do Vale
Lucelia Largura do Vale Vidigal
Marcos Tadeu Simões Piacentini
Talita Kelly Farias

SUMMARY

Robotics is a science that involves multidisciplinary concepts. It is now known that the new generation has great access to information, generating a lot of knowledge. As such, its use is very important for stimulating learning in children and young people. To this end, schools must adapt to the use of this science in order to improve the teaching and learning process. This research aims to identify the challenges of using robotics in a technical school in Cacoal, Rondônia. The research is classified as descriptive, qualitative, with a deductive method. To this end, a bibliographic survey was carried out in the first stage of the research. Based on this survey, a semi-structured interview was carried out with a teacher of the subject and a questionnaire with closed questions for 17 students in the subject of logic in the digital games course at the X teaching centre. In accordance with ethical considerations, the names of the students and the teacher will not be disclosed. The data collected showed that despite the challenges encountered by the teacher, the use of robotics as a teaching method contributes to achieving the students' learning objectives.

Keywords: Automation, robotics, education.

1 INTRODUCTION

Robotics is a technological area that involves disciplines such as mechanics, electronics, computing, automation and, above all, information systems. According to Ottoni (2010), the combination of these subjects makes it possible to create motorised mechanical systems that are controlled, either manually or automatically, by means of electrical circuits. Ribeiro (2001) states that the application of robots has become increasingly important in everyday life, as it allows their use to replace human or animal labour with the use of machines, Ottoni (2010) states that for this, it is necessary to invest in education focused on this area, in order to promote technological advances.

Educational robotics, also known as pedagogical robotics, is an area of robotics used to refer to education or learning. Santos (2012) says that this pedagogical method is characterised by the use of teachers, teaching materials and environments that provide learning through technologies such as robots. This learning can take place through the gathering of materials, whether scrap or assembly kits, made up of various parts, which can be computer programmed and the use of software to develop the functioning of the models initially proposed.

Santos (2012) states that models programmed using computers and software enable students to develop their learning autonomy, as well as stimulating logical reasoning, exercising creativity and experiencing previously taught content in practice. According to Lieberknecht (2009), using this method for learning can increase students' interest in the subject, as lessons can become more dynamic, arousing students' attention to the subject.

Santos (2012) says that these objectives should bring together the knowledge acquired previously in theory with the practice of solving the proposed problem. The teacher sets the same objective for several study groups and provides limited help in solving the problem. This assistance is limited in order to encourage students to develop their logical reasoning and decision-making autonomy. Group work provides students with a vision of teamwork, promoting dialogue and respect for different opinions. For Abreu et al (2005), teamwork provides an integration of individual skills and talents, transforming them into collective ones in order to produce more efficiently. Santos (2012) states that in order to develop these activities, a set of materials is needed, which can be robotics kits.

There are several different assembly kits for use in robotics that can be used to carry out teaching activities, but according to Santos (2013), the Lego kit (leg godt, which means "play well") stands out from other kits because it is a material that combines different parts that can be connected to each other and promotes the construction of robots or automated equipment, such as industrial equipment, depending on the objective the student wants to achieve.

As it is an interdisciplinary tool, the use of Lego can be used in various disciplines, making the content taught by the teacher more dynamic. According to Santos (2013), using the tool motivates students to learn the subject in question in a more practical way. An example of this is the use of Lego in physics to explain concepts of kinematics or rectilinear movements. The challenge proposed by the teacher can be to build simple or even complex robots.

The aim of the study was to raise the challenges encountered in using robotics in the teaching and learning process, surveying the advantages and disadvantages of its use, by means of field research using interviews and a closed questionnaire, applied to the teacher and students involved in the research. This research will strengthen the understanding and importance of using robotics as a means of learning, bearing in mind that advances in this area will provide a greater benefit to society, because the more widespread the subject of robotics, the greater progress will be made.

2 THEORETICAL FRAMEWORK

2.1 Automation

According to Silveira and Santos (2013), automation is defined as a group of methods for building systems that will act based on the information received and in an efficient manner. It is a feedback system, where the information is put into the system and the system in turn calculates the best way to carry out the action intended for it, it has an input/output relationship that can be corrected as the output does not correspond to what was desired.

Knowing that some dictionaries use the term automation as a synonym Silveira and Santos (2013), explain that it is related to the movement of machines, something that is repetitive, where there is no input/output relationship, meaning that there is no possibility of correction, as the system will always have the same behaviour.

For Silveira and Santos (2013), automation is associated with the concept of software, as it is more flexible, and automation is related to hardware, as it is more mechanical. Ribeiro (2001) states that automation is replacing work previously done completely by human beings with machines, where the human being has a small part to play, just adding intelligence to the machine. Alves (2005) states that automation is a technology in which three systems are connected: mechanical, electrical and electronic and controlled by a computer.

Automation has led to the invention of devices known as robots, which according to Ferreira (2001) are mechanisms that reproduce human activities and movements using a computer. Today they are widely used to help human beings carry out tasks in order to improve human life, whether in the professional or personal sphere.

2.2 Information and Communication Technology

According to Ramos (2008), information and communication technologies, known as ICTs, aim to process information and communicate it through procedures, methods and equipment. The IBGE - Brazilian Institute of Geography and Statistics - (2009), conceptualises ICTs as an arrangement of activities in industry, commerce and services which, in addition to transmitting and propagating data and information captured electronically, commercialise products specific to this medium.

According to Ramos (2008), they emerged against the backdrop of the Information Technology Revolution, also known as the Third Industrial Revolution, and enabled greater dynamism in the transmission of information and communication, since with them content was multiplied in images, digitalisations, videos and sounds, often united in a single product, thus providing three areas of application: control and automation, computers and communication, and within these they were subdivided into others, as can be seen below.

In the area of control and automation, Ramos (2008) subdivides it into robotics and CAD/CAM (Computer-Aided Design/Computer-Aided Manufacturing). CAD, which stands for computer-aided design, is software that makes it easy to create industrial, construction, industrial design and advertising products. CAM, which is the same as computer-aided manufacturing, means that all machine work is controlled by a computer. Robotics are systems made up of machines and mechanical elements controlled by computers.

Ramos (2008) divides the computer area into: robotics and informatics, where the latter is information passed on by means of automatic instruments, such as computers, and the former is the use of these informatics instruments in office environments. And lastly, in the area of communication, Ramos (2008) classifies it into: telecommunications and telematics, and says that in telecommunications the main objective is distance communication, whether by telephone, computer and so on, while in telematics, which is the combination of telecommunication and computing, the internet is used, aided by its own software for communication.

The IBGE (2009) shows that in 2003 ICTs became an instrument for economic, social and cultural development, and in 2004 digital inclusion became one of the country's priorities in order to have more information and knowledge. Chitolina and Scheid (2015) state that for the current generation of students, ICTs have become indispensable for generating knowledge, as they are as easy to access as the speed with which students seek this knowledge. According to Valente (1999), it is through computers that ICTs are being brought closer to students in the learning process.

2.3 Robotics

Robotics is the science that studies the assembly and programming of robots. Robots can be characterised as autonomous, reprogrammable devices controlled by software. The act of building and programming a robot requires the combination of knowledge from various fields, which gives robotics a multidisciplinary character. (LIEBERKNECHT, 2009)

As mentioned, robotics uses various sciences, where the knowledge of each area is applied so that the robot can be built. These sciences include: microelectronics (there are electronic parts in the robot), computing (the use of software and programming logic), mechanical engineering (the robot's mechanical parts), physics (the robot's movements) and artificial intelligence.

According to Rosário (2005), the term "robot" originated from the Czech word *"robotnik"*, which means "servant". Previously there was an idea that robots imitated man in their activities, referring to the idea of a "mechanical employee". In the 20th century, after the industrial revolution, there was a need to increase industrial productivity and product quality, at which time the first industrial robots began to appear.

Robots are useful in activities where there is a risk to human life or in operations where a very high degree of precision is required. They have applications in other areas, Santos (2013) points out that they are not only used in industrial activities, but also in non-industrial environments (such as domestic use, entertainment, police, etc.), medical (high-precision surgery), space (satellite maintenance, discovery of new planets), humanoid (have human characteristics), anthropomorphic (have human characteristics and those of other natural beings), virtual (do not exist physically).

For Lieberknecht (2009), the characteristics of robotics allow activities to become more productive when worked on in groups. This allows robotics to be a tool to aid teaching, as an innovative method for transmitting knowledge, also known as educational robotics.

2.4 Using Robotics in Education

Educational robotics has seen great growth in recent years, as it is a method that promotes the study of different contents in an attractive way, linking theory and practice, allowing students to solve the problems proposed by the teacher. According to Ragazzi (2007, apud Santos, 2013), the aim of educational robotics is to "get students to discover how technology works in a fun way", so robotics can also discuss accumulated knowledge and help students to "use, master and develop critical thinking". The individual development of the student also takes place in subjects that are not covered by a specific discipline, such as: teamwork, the ability to make decisions, solve challenges, critical thinking, creativity, logical reasoning, responsibility, among others.

Educational robotics aims to prepare students to develop their skills using assembly kits in learning environments. According to Lambert and Zilli (2010), the best-known and most widely used kit is the LEGO kit (leg godt, meaning "to play well"), which stands out from other kits because there is a wide variety of parts that make it up and it allows structures to be assembled flexibly, simply and quickly, such as industrial equipment, depending on the objective proposed by the teacher and that the student must achieve.

Conchinha (2011) emphasises that LEGO can be used by people who are new to electronics, as it is a set of systems with very simple electrical connections and fittings. Rarely has a child from an industrialised country not had some kind of contact with the toy in its initial version. For this reason, the author states that LEGO, when used in education, brings much more than leisure or excitement to students, but provides resources capable of attracting and developing skills.

2.4.1 Methods of using robotics in education

According to Moran (2013), the school, which is very adept at the conventional teaching system, has, even today, resisted changes in the learning method, in which the teacher is the sole holder of knowledge, and this tells us that there is still a long way to go. Moran (2013) also states that educators currently need to focus on influencing the construction of knowledge by the students themselves, so that the teaching generated in the institution is both collective, students/students, and personal, student/teacher.

According to Valente (1999), teaching is a service and, like everything else in society, changes occur as time goes by. And with these transformations have come new ways of transmitting knowledge to the student, of instigating them to seek knowledge and then perform better.

Thus, it is possible to find various robotic materials and ICTs that help develop students' learning skills. According to Valente (1999), the pedagogical methods to help students learn are: tutorials (software with a pre-organised pedagogical sequence to transmit information chosen by the student), programming (a tool that uses concepts and techniques to solve a problem), word processing (an application used to explain text in the person's natural language), multimedia and the internet (similar to tutorials, but with the use of images, sounds, etc.), developing multimedia or web pages (with this you have to search for information and thus knowledge is generated).), development of multimedia or web pages (with this you have to search for information and so knowledge is generated), simulation and modelling (in this the learner creates the phenomenon where you can change parameters to find out the result of this interference, and that the phenomenon is already ready by simply placing it on the computer) and games (challenge the learner by making them compete with other colleagues or with the machine itself).

According to Chitolina and Scheid (2015), in order to encourage the use of robotics in schools, Lego in partnership with Zoom created an educational programme with handouts, assembly kits and software. According to Feitosa (2013), LEGO offers various products ranging from toys for children to kits made up of sensors, motors and a central controller that manages the system, which is where the software is located. These kits come with various parts for assembling the robot.

Therefore, for Chitolina and Scheid (2015), robotics is accessible and using it for educational purposes allows students to build knowledge based on what they have learnt during their contact with the objects. Chitolina and Scheid (2015) apud Papert (2008), also say that with resources that provoke students' interest, such as robots and machines, learning and teaching become more pleasurable and rewarding.

2.4.2 Advantages of using robotics in education

The purpose of educational robotics is to provide students with the conditions to develop their skills. For Bezerra, Nascimento and Santos (2010), the main advantages are:

1. Allow learning to become motivating;
2. Provides autonomy in the search for knowledge;
3. Use simulators that are very similar to "reality";
4. It enables you to develop your ability to work in groups;
5. Increases the ability to argue and counter-argue;
6. Develops concentration and problem-solving skills;
7. Stimulates creativity;
8. Develops logical thinking;
9. It favours interdisciplinarity, since the student must receive knowledge from different areas.

Santos (2013) adds that educational robotics allows students to test knowledge previously learnt in theory on physical equipment. For Santos (2013), the use of robotics in education makes learning easier, as the method allows students to experience a given situation and be able to overcome the challenge proposed by the teacher.

2.4.3 Disadvantages of using robotics in education

According to Dantas Machado (2014) apud Vandevelde et. al (2013), the disadvantage is the high cost, making it impossible for people and institutions to afford such a price, another disadvantage is the limitation of programmable sensors and motors in each device, in addition it is very difficult to add other devices to the robot that are not from Lego itself and when possible there is damage to the Lego components.

According to Rosário (2005), what determines the price of a robot is its level of sophistication/technology; its size, the larger the more expensive; its complexity; the accuracy with which it works and the reliability it provides.

3 . RESULTS AND DISCUSSIONS

The research was carried out in a technical education institution X in the city of Cacoal, Rondônia. It has been operating in Brazil for over 70 years and in Rondônia for 55 years. It offers a range of courses in the areas of professional development, industrial apprenticeships, professional qualification and high-level technical training. The state of Rondônia currently has 7 (seven) institutions spread between the capital and the interior.

Data collection took place between 30 October 2015 and 06 November 2015 at the X technical education centre, located in the city of Cacoal- RO. The participating students are enrolled in the digital games technical course and robotics is used as a teaching method in the logic subject.

As a teaching method, the teacher uses a LEGO robotics kit containing parts for assembly, motors, sensors and software for building a robot, along with a handout containing challenges set for the students during the course. The students work in teams, where there is a programmer, a separator, an assembler and a leader, as shown in the image below:

The data was collected in two stages: an open interview with the teacher and a questionnaire with closed questions for the students. In the first stage, an open-ended interview was conducted with the teacher, where he answered questions about the methodology used, the aims of the subject, the challenges he faced with this methodology, and then he was asked to nominate three students to answer a closed questionnaire in order to test it. Since the questionnaire needed to be adapted, it was reworked and then applied to all the students, in total (17) seventeen, including those used in the test. The students' ages ranged from fifteen to eighteen. The questionnaire produced the following results:

GRAPH 1: ADVANTAGES OF ROBOTICS IN EDUCATION

SOURCE: PREPARED BY THE AUTHORS 2016.

When the students were asked about the advantages of using robotics for the course, 41.2 per cent of the results were for the development of logical reasoning, 29.4 per cent for expanding knowledge and 23.5 per cent for developing teamwork. As pointed out by Bezerra, Nascimento and Santos (2010) in topic 2.4.2 on the advantages of using robotics in education.

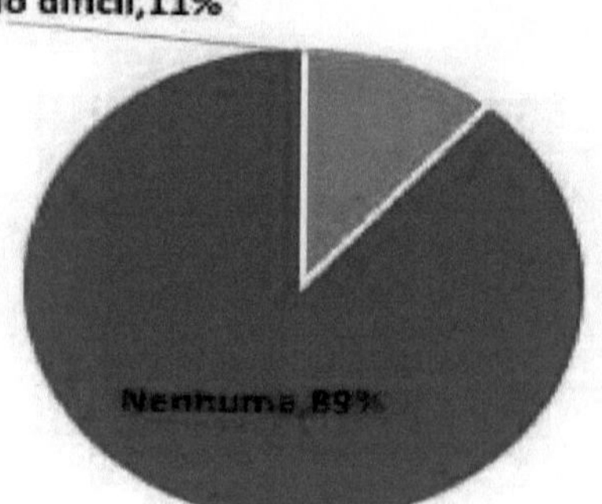

GRAPH 2: DISADVANTAGES OF USING ROBOTICS IN EDUCATION

SOURCE: PREPARED BY THE AUTHORS

Graph 2 represents the disadvantages of using robotics as a teaching method, and shows that 88.23 per cent of the students found no disadvantages to point out.

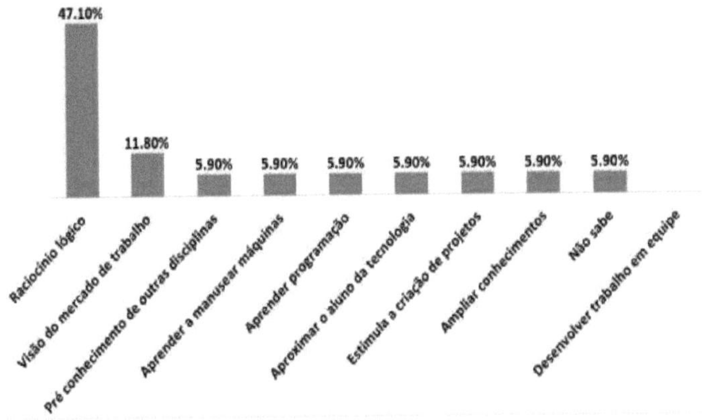

As shown in graph 3, according to the students' opinion, the main objective of the subject is to develop logical reasoning, with a result of 47.1 per cent, which confirms what Ragazzi (2007, apud Santos, 2013) stated in topic 2.4 of this work.

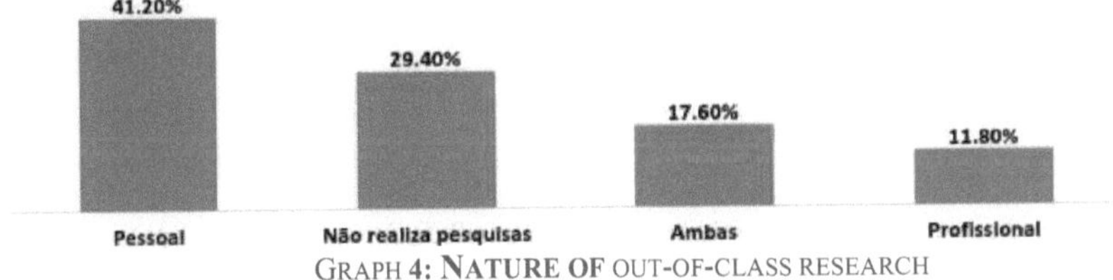

Graph 4 shows the nature of the out-of-class research into technology carried out by the students and showed a percentage of 41.2 per cent for personal research, 11.8 per cent for professional research and 17.6 per cent for both, with the remainder saying that they don't try to find out more about the subject.

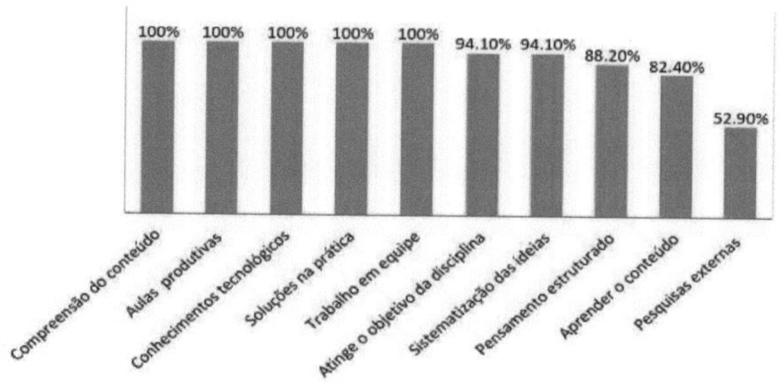

GRAPH 5: ADVANTAGES OF USING ROBOTICS

SOURCE: PREPARED BY THE AUTHORS.

In topic 2.4.2 on the advantages of using robotics in education, Bezerra, Nascimento and Santos (2010) pointed out some of the advantages of using robotics. This information is confirmed in Graph 5, where the students' opinion shows the positive points of using this resource. Where 100% of the students responded that the method has the following positive points: Facilitating understanding of the content; making lessons more productive; expanding technological knowledge; leading students to question and seek solutions, moving from theory to practice and facilitating learning and teamwork. With 94.1% it allows the student to achieve the subject's objective, followed by allowing the student to be challenged to think and systematise their ideas. 88.2% said that it favours the development of structured thinking; 82.4% indicated that it promotes greater interest in learning the content and 52.9% said that it promotes greater interest in research outside of classes.

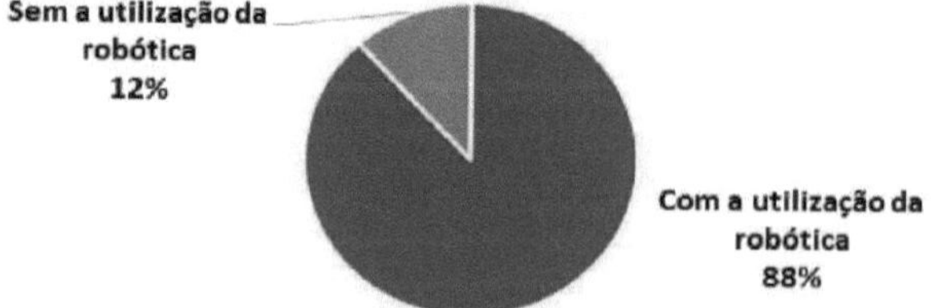

GRAPH 6: IMPACTS OF USING ROBOTICS IN MATHS SUBJECTS

SOURCE: PREPARED BY THE AUTHORS.

When asked about the use of robotics in exact subjects outside the course, the following result was obtained: 88.2% said that the use of robotics in these subjects had a greater positive impact on learning and 11.8% said that without robotics as a teaching method. This proves Lieberknecht's (2009) assertion about the multidisciplinarity of robotics.

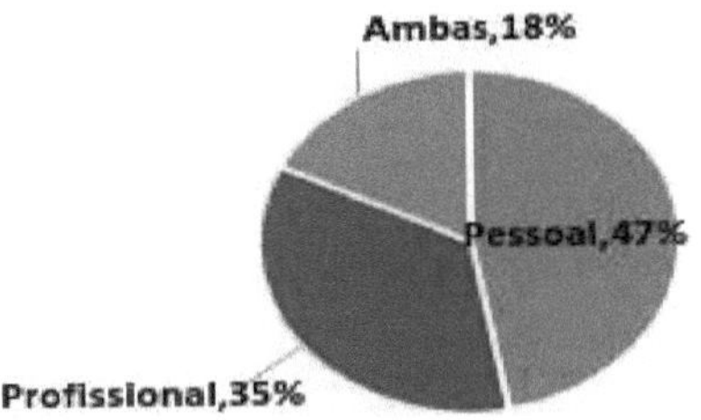

When asked why they took the course, the students replied: 47.1 per cent took it because they liked the subject; 35.3 per cent were looking for a job in the area and 17.6 per cent for both reasons.

The results collected from the students prove what the teacher said when asked about the method used, "the *use of robotics allows* students to develop skills such as logical reasoning, professional development, teamwork and makes learning the content more *dynamic."* The teacher said that he encourages students to develop their logical reasoning, leaving them independent when the challenge is launched, allowing them to ask questions, but not indicating a final answer, in order to encourage learning autonomy.

As a challenge to using the method, 11% of the students said that it was labour-intensive, but 89% said that there were no disadvantages to using it. The teacher of the subject says that one challenge encountered is the high cost of acquiring all the material, which is a disadvantage pointed out by Dantas Machado (2014) apud Vandevelde et. al (2013) in topic 2.4.3 of this work, which may be an obstacle to other teaching centres not adopting this methodology.

4 FINAL CONSIDERATIONS

As explained in the literature review, robotics is a science that involves multidisciplinary concepts and is currently part of teaching and learning methods in schools for children, young people and even adults. The aim of the research was to identify the challenges of using robotics in a technical school in Cacoal, Rondônia, where the advantages and disadvantages of the teaching method would be investigated.

The study was classified as descriptive, qualitative, with a deductive method. Data was collected through interviews with the teacher and closed questionnaires with the students. After analysing the data, we came to the conclusion that robotics has great importance in teaching didactics, enabling advantages with its use. The authors mentioned that robotics really does develop students' skills and provides better learning, as well as professional growth, preparing students for the labour market. Unfortunately, it has been little used as a teaching methodology because of the high costs of acquiring materials, so the challenge is to spread the idea of using this didactic method.

It has added to academic knowledge about the importance of research related to education, because once you know the challenges of using this form of learning, you can set goals to overcome them. It has brought the knowledge that this teaching method can improve the learning of many who need to use the skills provided by LEGO, such as the systematisation of ideas, logical reasoning and others. It is also important to publicise the method, so that it can facilitate the insertion of this way of transmitting knowledge to society, in order to achieve the goal of education.

6 REFERENCES

ABREU et al. Teamwork in nursing: a systematic literature review. **Brazilian Journal of Nursing.** Brasília. v58. n2. Apr. 2005. Available at: < http://www.scielo.br/scielo.php?pid=S0034-71672005000200015&script=sci_arttext > Accessed on 22 December 2015.

ALVES, T. S. **Industrial Automation I.** Abrantes. 2005. Available at: < http://www.ebah.com.br/content/ABAAAfiswAF/automacao-industrial > Accessed on 28 November 2015.

BEZERRA, R. M. S. NASCIMENTO, F. M. S. SANTOS, F. L. S. **Educational robotics as a low-cost approach to teaching computing in technical and technological courses.** Available at: <http://ceie-sbc.educacao.ws/WIE2010/pdf/st06_02.pdf> Accessed on 01 December 2015.

CHITOLINA, R. F.; SCHEID, Neusa Maria John. **Educational robotics and information and communication technologies in the construction of substantive knowledge in the natural sciences.** 2015. Available at: < http://cascavel.ufsm.br/revistas/ojs-2.2.2/index.php/cienciaenatura/article/viewFile/14834/pdf_1 > Accessed on 02 December 2015.

CONCHINHA, C. I. **Lego Mindstorms: a study with users with cerebral palsy.** Available at : <http://repositorio.ul.pt/bitstream/10451/5747/1/ulfpie039843_tm.pdf> Accessed 01 December 2015.

FEITOSA, Jefferson Gustavo. **Pedagogical teaching manual.** 1 ed. Curitiba: Zoom Editora Educacional, 2013.

FERREIRA, Aurélio B. de Hollanda. **Miniaurélio século XXI Escolar:** O Mini-dicionário da Língua

Portuguesa. 4. rev. ed. expanded. Rio de Janeiro: Nova Fronteira, 2001.

LAMBERT, G. ZILLI, G. M. **Developing education through mobile robotics: a pedagogical proposal for teaching engineering.** Available at

<http://www.abenge.org.br/CobengeAnteriores/2010/artigos/606.pdf> Accessed 01 December 2015.

LIEBERKNECHT, E. **Robótica educacional.** 07 August 2009. Available at <http://www.portalrobotica.com.br/site/index.php?option=com_content&view=categor y&layout=blog&id=5&Itemid=2> Accessed on 01 December 2015.

MORAN, José. The integration of technologies in education. Campinas: Papirus. 5 ed. 2013.

OTTONI, A. L. C. Study material: introduction to robotics. Available at <http://www.ufsj.edu.br/portal2-repositorio/File/orcv/materialdeestudo_introducaoarobotica.pdf> . Accessed on 09 December 2015.

RIBEIRO, M. A. **Industrial Automation.** Salvador. 4 ed. 2001. Available at: < http://eletro.g12.br/arquivos/materiais/eletronica6.pdf > Accessed on 25 November 2015.

ROSÁRIO, J.M. **Princípios de Mecatrônica.** São Paulo: Prentice Hall, 2005.

SANTOS, I. **What is educational robotics?** Available at: <http://www.roboticanaescola.com.br/> Accessed on 09 December 2015.

SANTOS, T . N. **The application of robotics in the teaching-learning process in basic education.** Federal University of Santa Catarina, Araranguá, December 2013
                         Available at   :
<https://repositorio.ufsc.br/bitstream/handle/123456789/131069/Tatiana_Santos.pdf? sequence=1&isAllowed=y>. Accessed on: 25 November 2015.

SILVEIRA, Paulo Rogério da; SANTOS Windersom E. dos. **Automation and Discrete Control.** São Paulo: Érica, 9 ed 2013.

VALENTE, José Armando. **The computer in the knowledge society.** Campinas.   1999.
                         Available at   :<
http://www.fe.unb.br/catedraunescoead/areas/menu/publicacoes/livros-de-interesse-

in the-area-of-tics-in-education/the-computer-in-the-society-of-knowledge >
Accessed on: 01 December 2015.

CHAPTER 4

THE CHALLENGES OF USING AUTOMATION IN A RECORD COMPANY

Ademir Luiz Vidigal Filho
André Jun Miki
Carlaile Largura do Vale
Maria Conceição de Paula Viera
Wesley Gonçalves Pereira

SUMMARY

This article is a descriptive study of the challenges of applying industrial automation in a record company, due to the technological innovations incorporated into its business segment. In short, it seeks to understand the effects of different communication technologies on the development of this sector. The aim is to analyse the challenges of using automation in the production of phonographic services, especially in relation to purchasing, entering the market and winning new clients. To this end, a semi-structured interview was carried out, with qualitative data and open and closed questions about the challenges faced by producers when using automation in the music industry.

Keywords: Phonographic industry. Automation. Internet.

1 INTRODUCTION

The music business in Brazil and around the world is going through a period of transition and reinvention like never before. The music industry has undergone so many changes since the turn of the century that some even think it no longer exists. Its business model, which used to be fundamentally based on physical media (vinyl, cassette tapes, CDs), has lost ground with the advance of a new digital complex, which, in turn, offers various possibilities for music consumption (FERREIRA, 2013).

The Academy of Marketing (2016) emphasises that companies' growing interest in digital media is creating a demand that simply couldn't be met previously by digital agencies. This type of market is expanding dramatically due to the growth of the economy in recent years, as well as the increase in e-commerce with a progressive adhesion of consumers from formal commerce.

Ferreira (2013) says that the reduction in record company revenues is usually explained by the influence of piracy in physical media (through the illegal trade in counterfeit media), but mainly in virtual media (through unauthorised exchanges of audio files). At the end of the 1990s, CD burning technology became accessible to home users, coupled with the emergence of the first internet file-sharing programme (Napster, in 1999), which boosted free music consumption irreversibly.

The Brazilian phonographic market is beginning to recover after suffering a decline in CD sales and is investing in new audio and video distribution formats on the internet. According to projections by Universal Music Brasil, the industry is expected to double in size by 2020, opening up opportunities for producers, linked to advances in automation (GUIA DO ESTUDANTE, 2016).

The 1970s were also an important time in the computing process. According to Lévy (1999), the development and commercialisation of the microprocessor (an arithmetic and logic calculation unit located on a small electronic chip) triggered several wide-ranging economic and social processes.

They ushered in a new phase in the automation of industrial production: robotics, flexible production lines, industrial machines with digital controls, etc. They also saw the beginning of automation in some tertiary sectors (banks, insurance companies). Since then, the systematic search for productivity gains through various forms of use of electronic devices, computers and data communication networks has gradually taken over all economic activities. This trend continues today (LÉVY, 1999, p. 31).

On the other hand, from this period onwards, the computer progressively escaped from the data processing services of large companies and professional programmers to gradually become a tool for creation, organisation, simulation and entertainment (games) in the hands of a growing proportion of the population in developed countries - as can be seen today.

2 THEORETICAL FRAMEWORK

2.1 Production Systems

According to Tubino (2007), in order for a production system to be able to transform inputs into products, it needs to be planned in terms of deadlines, in which actions are triggered based on production planning, according to the estimated deadlines for each production schedule, so that the events planned by the companies become a reality. In general terms, the planning horizon of a production system can be divided into three levels: the long term, the medium term and the short term. A production system will be all the more efficient if it manages to synchronise the transition from strategies to tactics and from tactics to production and sales operations for the products requested.

According to Tubino (2000), there are different ways of classifying a production system and these will therefore directly influence the way production planning and control is carried out. There are three different classifications. The first classification would be by the degree of standardisation of the products, where they are divided into: standardised products and made-to-measure products. Standardised products are those that are manufactured in large quantities and uniformly throughout the system and are always available to customers on the market, so the company finds it easy to standardise its working methods. Bespoke products, on the other hand, are those that are produced to a specific customer's order. There is some difficulty in standardising work procedures and these products are generally more expensive.

The second classification is by type of operation and can be divided into: continuous processes and discrete processes. Continuous processes can be defined by the existence of a high degree of uniformity in production, the system is not flexible and the processes are interdependent. Discrete processes are so called because the products can be isolated, individually or in batches. They can also be divided into: mass repetitive processes, batch repetitive processes and project processes. The mass repetitive process is used for standardised products in large quantities, is not very flexible and any changes are made during final assembly of the product. The batch repetitive process is defined as the manufacture of standardised products in certain quantities (batches) and the production system is flexible as the product changes. Project-based processes are characterised by fulfilling a certain customer's order, production is all geared towards this goal and is altered when a new project is received.

The third and final classification is by the nature of the product, which can be a good (tangible) or service (intangible). The big difference between them lies in the way the tasks are carried out. The production of a good will be product-orientated, while the production of a service will be action-orientated. In general, the type of production system adopted by the company, as Tubino (2000, p.31) explains, "directly implies the level of complexity required to carry out production planning and control".

Studies in operations management have highlighted the production system as something that must be worked on in a process of continuous improvement, based on optimising the production system. For this reason, companies are constantly seeking goals aimed at making organisations and processes more efficient, such as: quality, reducing costs, increasing productivity and guaranteeing customer loyalty by reducing delivery times.

Obtaining reliability with the established objectives is achieved by the company by planning its strategic actions. As such, PCP (Production Planning and Control) streamlines management support for production by indicating improvements in decision-making to achieve the company's desired effects.

2.1.1 Discrete production system

Saad (2012) states that any discrete production system is responsible for distinguishing a product. At the end of a discontinuous process, individual products can be identified from one another even if they are basically identical. This type of production is usually separated from the manufacturing process.

Discrete manufacturing does not produce a homogeneous output, it is based on having more flexibility in its processes due to their reversibility than manufacturing processes. Most discrete manufacturing products can be disassembled and returned to their original components, the smallest of these components probably being the result of a manufacturing process (SAAD, 2012).

The end products of discrete manufacturing can contain serial numbers on all products and be sold with individual price tags and barcodes. Examples of discrete manufacturing include cars, boats and aeroplanes. These items have high individual values and are consequently treated with great individual attention in the production process. As these products move along the assembly lines, their individual units gain more and more value. Lower-value items, such as household appliances or furniture, are also discrete manufacturing products, however, because they are individually separated (SAAD, 2012).

2.1.2 Intermittent production system

Paula (2008) states that intermittent production is carried out in repetitive processes in batches and repetitive processes to order. At the end of the manufacturing process for a batch of a particular product, other products take their place on the machines. The original product will only be manufactured again after some time has passed, thus characterising intermittent production of each product. For example, in metallurgical plants that divide operations into stages and on the same machine, the first process is carried out, the machine is stopped and production of the second process begins, when it is finished it returns to the first process.

According to Zacarelli (1979), the diversity of products manufactured and the small size of the manufacturing batch mean that the equipment is subject to frequent variations in work and is subdivided into the following:

Made-to-order manufacture of different products: product according to customer specifications and manufacture begins after the product has been sold;

Repetitive manufacturing of the same batches of products: products standardised by the manufacturer, repetitive manufacturing batches, you can have the same flow characteristics as in custom manufacturing.

Moreira (1998) defines intermittent production systems (intermittent flow) at the following levels:

By batches: when a product is finished being manufactured, other products take their place on the machines, so that the first product will not be manufactured again until some time has passed;

Made to order: the customer submits their own product design and these specifications must be followed in manufacture.

According to Tubino (2009), a production-to-order system aims to set up a production system to meet the specific needs of customers with low demands. The product has a specific date negotiated with the customer to be manufactured, and once completed, the production system turns to a new project. Products are designed in close liaison with customers, so that their specifications require an organisation dedicated to the project, which cannot be prepared in advance, especially with the generation of intermediate stocks to speed up production lead times.

2.2.3 Mass production system

Slack (1997) says that a mass production system produces goods in high volume and relatively narrow variety, i.e. in terms of the fundamental aspects of the product design. A car factory, for example, could produce several thousand car variants if all the options for engine size, colour, extra equipment, etc. were taken into account.

They have a linear sequence for producing products and are fairly standardised, flowing from one station to another in a predictable sequence. When the type of product processed is discrete, the production system is called mass manufacturing or repetitive manufacturing. If the type of product processed is continuous, as in the process industries (chemicals, paper, etc.), manufacturing is called continuous (SLACK, 1997).

2.2 Automation

Automation is a word directly linked to automatic control, i.e. actions carried out without human intervention. Therefore, this concept requires conceptual discussions in order to explain the "non-human intervention" in automated actions. The Houaiss Conciso dictionary (2011) defines automation as the automatic operation of mechanical or electronic devices; a process or system in place of human labour; the process of replacing human labour with this type of operation.

Aiming to simplify man's work by interconnecting and/or replacing manual labour with mechanisms and means of control, the agility of the processes made it possible to have more time available to carry out other tasks or even to enjoy the arts and leisure. This simplification of work came about with the use of the wheel between 3500 and 3200 BC, linking mechanisation with automation (SILVEIRA and SANTOS, 1998).

According to Parede (2011), automation is an area that is gradually expanding its activities. The use of devices and the application of solutions are not just a substitute for human labour, they bring improvements in the quality of processes, optimisation of spaces, reduction in production time and costs. According to Rosário (2005), industrial automation can be understood as a technology that integrates the following areas:

a. Electronics: responsible for the hardware;
b. Mechanical: in the form of mechanical devices (actuators);
c. IT: responsible for the software that will control the entire system.

Automated systems can be applied to simple machines or to the entire industry. Nowadays, industrial automation is widely used to improve productivity and quality in processes that are considered repetitive, and is present in companies' day-to-day operations to support production concepts such as the Toyota Production System.

2.2.1 Types of automation

From a production point of view, industrial automation can be divided into three types: fixed automation, programmable automation and flexible automation.

Fixed automation consists of specialist workstations that process the product in such a way that they specialise in a particular task, specific to a particular type of product. This process is used when the volume of production is very high and, because the useful life of a product is compromised, the specialised machine becomes outdated. The fixed automation system is an essentially automatic process, which may or may not require an operator.

Programmable automation has product adaptability characteristics, i.e. it makes the process capable of being reprogrammed when the technical specifications for manufacturing an artefact undergo any kind of change. This characteristic is realised when the production volume is relatively low and there is a variety of products to be manufactured. In programmable automation, when a batch is completed, the equipment is reprogrammed for the next batch.

Flexible automation combines characteristics of programmable and fixed automation, forming an intermediate type in which flexibility is the fact that several types of product can be manufactured at the same time within the same manufacturing system. It is used for medium production quantities. The flexible system allows changes to be made at any time.

2.2.2   Advantages and disadvantages of automation

Silveira and Santos (1998) highlight the following advantages of automation for the production process:

a.   This is a process of irreversible technological evolution;

b.   Valuing human beings when they are freed from tedious and repetitive tasks, or even unhealthy and risky work situations;

c.   Increased quality of life for society as a whole, promoting comfort and greater integration;

d.   Greater enrichment through lower product costs or increased productivity;

e.   A question of survival and a strong marketing appeal in a highly competitive market;

f.   Creation of direct and indirect jobs, as well as new jobs related to the maintenance, development and supervision of systems;

g.   Striving for product quality and customer satisfaction.

Among the disadvantages, Silveira and Santos (1998) highlight the following: because it is irreversible, it becomes unpredictable, the consequences of which can only be assessed in the future; because it requires an increasingly qualified professional to carry out these functions, it has imposed a tapering employment policy; as the population starts to grow in a disorderly fashion, social inequalities increase, causing a limit of instability; as with all new technology, it can bring serious risks to the production sector; due to the globalised market, only large groups of companies have quick and easy access to all this new technology; there is an immediate reduction in available jobs; in the search for quality, craftsmen are no longer valued, but rather large-scale production, making people increasingly technologically dependent.

The benefits offered by industrial automation are huge. For many years, researchers have endeavoured to develop highly technological solutions. Acquisition costs have fallen and alternatives have increased, so more and more different industries have started to implement industrial automation systems to speed up production capacity. The benefits of industrial automation have become very evident, which has made automation a more attractive alternative.

3 METHODOLOGY

This research in the form of an article was carried out descriptively with the aim of analysing the challenges of using automation in the production of phonographic services, in which qualitative data was observed. The main objectives of this research are to present the concepts of production systems and automation and to identify the production system used in the organisation; to identify these automation challenges for the production company and to analyse them.

In the first stage of the research, a bibliographical search was carried out, mainly in books and articles, with the aim of conceptualising the subject, where the information collected was analysed, identified and put into a simple and objective form.

In the second stage, a semi-structured interview was carried out with open questions based on the bibliographical research, in which information on the proposed topic was collected. The interview with open and closed questions was conducted with the producer of a phonographic company located in the municipality of Vilhena, in the state of Rondônia. The research will follow the ethical aspects related to the bibliographic sources used in the study, respecting the integrity of the information and the confidentiality of the research participant.

In the third stage, the data was compiled in order to analyse the information obtained through the interview. The data from the interview applied to the record producer was analysed and selected in order to better understand the answers to the objectives to be achieved in this research.

4 ANALYSING RESULTS

This study was carried out on the basis of occurrences of a set of characteristics by obtaining data and analysing the challenges of using automation in the production company. The interview was carried out with the producer at a record company located in the municipality of Vilhena, in Rondônia, with open and closed questions, making it possible to analyse and understand the importance of automation for business.

4.1 Historical Context

The company "Áudio Sound Produtora" has been in the phonographic market since 2014, located in Vilhena, in the state of Rondônia. The production company makes recordings and produces texts, spots, vignettes, waits, political campaigns, material for rodeos and any other recording the client requests.

After working for a few radio stations in the city as an announcer and vignette producer, the owner became interested in the music production sector and even though he was away for a long time, working in other sectors such as commercial sales, pharmacy, among others, there was always the desire to set up an audio production studio. In June 2014, the company Áudio Sound Produtora was founded to serve radio stations, TV stations, advertising agencies and/or other audio production companies.

The creation and recording work is done in the studio, but the presentation, negotiation and delivery of the material is done 100% virtually, not restricting the use of telephone calls for negotiations and possible client acquisitions, thus serving the entire national territory, with the possibility of overcoming geographical barriers, such as the making of corporate audio (videos presented at business meetings, training sessions, product presentations, etc.). In addition to audio, the company also produces campaigns, texts and outsources services such as making videos, jingles (advertising in the form of music) and vignettes for radio stations.

4.2 Results

The interview took place at the company "Áudio Sound Produtora", with the owner, where 15 open and closed questions were asked, the results of which are shown below.

The organisation is in the intermittent production system (to order), as mentioned by Paula (2008), in which the company only starts the production process once the order has been received. On receiving the order, the customer is given a quote which, if approved, will serve as the basis for what will be produced. The organisation's marketing campaign strategy is based on winning over customers via e-mail and phone calls to publicise its work, as well as the company's website.

With the advance of the internet, the company Áudio Sound Produtora highlights some advantages, such as: a greater reach of clients, speedy delivery of the final product, as well as reaching a greater number of clients and being able to serve the entire national territory. Among the disadvantages, according to the owner, is the increase in competition, as there is a growth in professionals competing for the same clients, and the devaluation of the service, due to the presence of unqualified professionals who practise unfair competition.

The breaking down of geographical barriers due to the advance of the internet and automation brings benefits and agility to audio production. In a verbalised extract, the interviewee states:

"I don't see a ceiling to the benefits of breaking down geographical barriers, with the internet you can serve the whole world, you just have to be qualified. Breaking down geographical barriers is beneficial for audio production up to the point where you can serve the client with quality, since the internet, for example, would be the end point of the service." (verbal information).

The equipment used in a record production company generally requires high costs. High investment in equipment must be analysed beforehand in order to reduce these essential costs. The interviewee emphasises that every investment must be assessed in advance, because there is equipment that is viable to purchase, and other equipment that is only viable for large production companies.

The phonographic market is growing, according to the website Guia do Estudante (2016), following the crisis. According to the interviewee's verbalised extract, "there's always room for creativity, as long as the client's investment doesn't have to be more than they're willing to pay".

Internet platforms (emails, websites) are essential for reaching the target audience, as the production company has a large number of clients outside its local area, as well as making it possible to publicise its projects, send portfolios and materials in order to reach new clients.

According to the interviewee, the company Áudio Sound Produtora bases its marketing and publicity on its website, providing future clients with a bank of voices, containing a variety of options to choose from, as well as portfolios of speakers and advertising campaigns. The client relationship strategy can also be achieved through social networks, depending on the openness of each client; we use social networks to create bonds of proximity.

5 CONCLUSION

This article sought to highlight the challenges that an organisation faces when using automation, applying the theme to a recording company located in the municipality of Vilhena, in the state of Rondônia. The results obtained with this descriptive research adopted questions that sought to question the subject addressed in order to study the company "Áudio Sound Produtora" and it was found that it remains in the market always looking for improvements, aiming to reach customers, either through e-mails or even advertising campaigns.

Technological advances combined with the use of the internet have helped the company to reach an effective audience, breaking down the geographical barriers that many companies are limited to. For the music industry, the use of automation is essential, as it is difficult to store the bank of voices that the announcers generate for the companies that hire this service, and no less important to seek and maintain a relationship between client and company (via websites, emails and campaigns).

The application of automation is fundamental in today's market, especially with the aim of creating a competitive advantage that allows the company to survive in fierce/competitive markets. It is worth emphasising that the flexibility of production, whether of goods or services, allows companies to explore common market niches in order to increase their share of consumers.

This article is important for the academic community who are looking for basic concepts in the literature that go towards knowledge. It is suggested that this article be complemented by a study of the advantages and disadvantages of geographical barriers for the music industry, and how these barriers influence the organisation's productivity and dissemination.

REFERENCES

Available at   :
<http://www.academiadomarketing.com.br/mercado-de-trabalho-marketing- digital/>. Accessed on: 08 June 2016.

Áudio Sound Produtora. Available at: <http://www.audiosound.com.br/>. Accessed on: 20 May 2016.

CAPELLI, Alexandre. Industrial Automation: motion control and continuous processes/Alexandre Capelli. 2ª ed. São Paulo: Érica, 2008.

CONCISO, Houaiss. Houaiss Concise Dictionary. São Paulo: Editora Moderna. 2011.

FERREIRA, João Miguel C. "Phonographic industry and the new sound platforms: an exploratory study on the Brazilian music market". Brasília, 2013.

Available at   :
<http://guiadoestudante.abril.com.br/profissoes/comunicacao-informacao/producao-   fonografica-687260.shtml>. Accessed on: 08 June 2016.

LÉVY, Pierre. Cyberculture. São Paulo: Ed. 34, 1999.

MOREIRA, Daniel A. Production and Operations Management. 3. ed. São Paulo: Pioneira, 1998.

PAREDE, Ismael Moura. Electronics: Industrial Automation / Ismael Moura Parede, Luiz Eduardo Lemes Gomes (authors); Edson Horta (co-author), Luiz Carlos da Cunha e Silva (reviewer); Jun Suzuki (coordinator). -- São Paulo: Fundação Padre Anchieta, 2011 (Coleção Técnica Interativa. Série Eletrônica, v. 6).

PAULA, Wagner de. Production Management. Published on 11 June 2008. Available at: <http://www.administradores.com.br/artigos/carreira/a-administracao-   da-producao/23401/>. Accessed on: 11 May 2016.

ROSÁRIO, J. Principles of Mechatronics. São Paulo: Pearson. 2005.

SAAD, Flávia. What is discreet manufacturing? Available at: http://www.manutencaoesuprimentos.com.br/conteudo/7138-o-que-e-fabricacao-   discrete/. Published on 11 September 2012. Accessed on: 01 May 2016.

SILVEIRA, Paulo Rogério da, 1968. Automation and discrete control 9 ed./Paulo Rogério da Silveira, Winderson E. Santos. São Paulo: Érica, 1998.

SLACK, N., Chambers, S., Harland, C., Harrison, A. & Johnston, R. "Production Management". São Paulo: Ed. Atlas (1997).

TUBINO, Dalvio Ferrari. Manual de planejamento e controle da produção. 2. ed. São Paulo: Atlas, 2000.

TUBINO, Dalvio Ferrari. Production Planning and Control Manual. São Paulo: Atlas, 2009.

ZACARELLI, Sérgio Baptista. Production Programming and Control. 5. ed. São Paulo: Pioneira, 1979.

CHAPTER 5

THE BENEFITS OF PRODUCTION AUTOMATION FOR WORKERS IN A PLANNED FURNITURE FACTORY IN THE MUNICIPALITY OF CACOAL

André Grecco Carvalho
André Jun Miki
Denny William de Oliveira Mesquita
Cristina dos Santos
Tauane Karine Souza Nascimento

SUMMARY

In recent years, countless social, political, economic and technological changes have been taking place, making significant modifications to production sectors necessary. The furniture industry can be considered one of the oldest in the world, as it derives from the carpenters and craftsmen who made furniture. With the industrial revolution, they began to use machines and tools in order to save effort and time. The furniture industry can be classified as a traditional industry, with consolidated and widespread production technology. The aim is to identify the benefits of production automation for workers in a planned furniture factory in the municipality of Cacoal, with the specific objectives of identifying the organisation's production system, describing the advantages and disadvantages of production automation, collecting quantitative data on the history of accidents and identifying the benefits of automation for workers. This research is classified as descriptive, with a qualitative approach and the method used was deductive. The research used bibliographical references obtained from books, scientific articles and dissertations in the area under study and, as a result, it was possible to generate a semi-structured interview applied to the company's production manager and a questionnaire applied to seven employees whose activities are directly linked to the automated sectors of the Bianchini furniture company. The data collected and analysed shows that the process of automating the company in question was a lengthy one with a high investment value, and one of the biggest obstacles to achieving the desired objectives is the resistance of certain employees to answering correctly. The questionnaire was administered between the nineteenth and twenty-fifth of May in the year two thousand and sixteen.

Keywords: Automation, furniture industry, occupational safety.

1 INTRODUCTION

Man has always sought simpler, faster and more precise ways of doing his work. This can be seen in the development and creation of tools in the stone age, passing through various other inventions and culminating in the industrial revolution, a major milestone in which machines came to replace human labour once and for all. With it came mass production, the production line and many other concepts that are part of our daily lives (OLIVEIRA, 2015).

The furniture industry can be considered one of the oldest in the world, as it derives from the carpenters and craftsmen who made furniture. With the industrial revolution, they began to use machines and tools in order to save effort and time. The advances brought about by industrialisation allowed for standardisation and gains in scale, so that furniture ceased to be handmade products and became industrialised (FERREIRA et al, 2008).

There has been significant growth in the furniture sector in Brazil in recent years. Data from the Brazilian Furniture Industry Association (ABIMÓVEL) in 2007 shows that the Brazilian furniture industry generates more than 206,000 jobs, distributed among approximately 16,000 micro, small and medium-sized companies (GOMES and GUIZZE, 2015).

The furniture industry can be classified as a traditional industry, with consolidated and widespread production technology. In turn, the technological dynamism of this industry is determined by the improvement of design, the machinery and equipment used in the production process and the introduction of new materials. Design is the only innovation factor specific to the furniture industry and is linked to all the aspects related to product design that allow the company to differentiate itself and build advantages over competitors: manufacturability, ergonomics, quality, durability, comfort, use of new materials, distribution and marketing strategies, among others, providing a company's furniture with its own identity (TIGRE, 2006).

According to updated data from IEMI (Market Intelligence, 2014), the country has 18,200 production units operating in the furniture sector, employing 300,000 people, an increase of 28 per cent compared to 2009, from 237,000 to 303,000 people employed directly and indirectly in the sector (GOMES and GUIZZE, 2015).

Automatic machines have brought new handling and control requirements and new qualification demands. Both in academic studies and in the field of administration and business, the subject of quality has gained ground and importance.

Small or family-run companies have always sought to make furniture to order. With technological developments, automation has become more accessible and advantageous for this market, especially in terms of cutting tools and design, which makes it easier to manufacture this furniture. Because it has always been an activity that required a great deal of physical effort, with the arrival of automation there has been a considerable reduction in the effort that the worker would make in the manufacture of a particular piece of furniture (CARVALHO; SOUZA and LIBOREIRO, 2006).

Automation has brought countless benefits to companies and workers, making it easier to manufacture furniture in less time and at a lower cost, reducing physical effort on the part of workers and making it safer for them to carry out their activities.

2 THEORETICAL FRAMEWORK

This topic will present some characteristics of production systems, occupational safety, the history of automation, the advantages of using industrial automation in the furniture sector, and the disadvantages found in the automation process. With the aim of developing ideas based on bibliographical references guiding the research, presenting a foundation of literature already published on the same subject.

2.1 Production Systems

Slack, Chambers and Johnston (2002) explain that any operation that produces goods or services, or a mixture of the two, does so through a process of transformation and, by transformation, they mean the use of resources to change the state or condition of something in order to produce outputs. In short, production involves a set of input resources used to transform something or to be transformed into outputs of goods or services.

Moreira (2002) explains that a 'production system' is the set of interrelated operational activities involved in the production of goods (in the case of industries) or services. The production system is still an abstract entity, but it is extremely useful for giving an idea of totality.

Tubino (1997, p.27) discusses the classifications of production systems more broadly and identifies the criteria on which three of them are based:

1. By degree of standardisation: Systems that produce standardised products such as goods or services that have a high degree of uniformity and are produced on a large scale, and systems that produce tailor-made products: goods or services developed for a specific client.

2. By type of operation: Continuous processes involve the production of goods or services that cannot be identified individually. Discrete processes involve the production of goods or services that cannot be isolated, in batches or units, and identified in relation to others. They can be subdivided into Mass repetitive processes: large-scale production of highly standardised products; batch repetitive processes: production in batches of an average volume of standardised goods or services; project-based processes: meeting a specific customer need, the product designed in close liaison with the customer has a set date for completion. Once completed, the production system turns to a new project.

3. By the nature of the product: Manufacturing of goods: when the product manufactured is tangible; service provider when the product generated is intangible.

A Flexible System is a grouping of semi-independent computer-controlled workstations linked by an automated transport (or handling) system. Its implementation is recommended when there is a high variety of parts to be produced, in low and medium production volumes (MARTINS, 2002).

2.2 Automation

Since prehistoric times, human beings have somehow processed stones, then metals, then more and more elaborate pieces until they reached the construction of simple and efficient machines, but with manual propulsion. That's why they weren't yet considered machine tools, machines capable of extending intelligent human action without their own energy (ABIMAQ, 2006).

The concept of automation is defined as the technique of making a process or system automatic, referring to both services performed and products manufactured automatically (BLACK, 1998).

According to Silveira (2011), automation is defined as the combination of the tools needed to produce a particular item, with little or no intervention by human labour. An example of this is a continuous flow production process that involves various mechanisms to produce an item. In an industrial context, automation can be defined as the technology that deals with the use of mechanical, electro-electronic and computerised systems in the operation and control of production (PAZOS, 2002).

When automating a production process, it is necessary to use mechanical, electrical and electronic devices that perform functions equivalent to those of humans in supervision and control activities, such as collecting and analysing data and correcting course (GUTIERREZ; PAN, 2008).

Automatisation and automation have different concepts, with automatisation being inextricably linked to the suggestion of automatic, repetitive, mechanical movement and thus synonymous with mechanisation, i.e. a mechanism of blind, uncorrected action. However, automation is a set of techniques through which active systems are built that are capable of acting with optimum efficiency through the use of information received from the environment on which they act (SANTOS, 1979).

For GROOVER (2001), automation can be defined as a technology concerned with the application of mechanics, electronics and computer-based systems to operate and control production.

The automated elements of a production system can be divided into two categories:

a. Automating the manufacturing system in a factory;
b. Computerisation of the manufacturing support system.

Automated manufacturing systems can also be categorised into three basic types: fixed automation, programmable automation and flexible automation.

Automation is referred to in four areas: automation of machines for processing, of materials handling, of controls and of design, with the merger of processing and controls providing a great deal of scope for development (ANTUNES, 1996).

According to LUZ and KUIAWINSKI (2006), in automation the control is carried out by the machine with the assistance of an operator, and these automated machines are not equipped with a human brain. The system is able to calculate the most appropriate corrective action. It is used to complete the integration of the STP (Toyota Production System) from an automated operation - it is an activity to improve integration and flexibility in a production process.

2.2.1 Advantages of automation in the furniture sector

Industrial automation is a way that many companies have found to improve the production process of their products. One of the advantages of using industrial automation is the fact that machines, combined with technological advances and information technology, can do a man's job better and faster (PORTAL EDUCAÇÃO, 2016).

The machinery and equipment used by this industry has undergone a major change in recent decades, with the electromechanical base being replaced by microelectronics, allowing for greater use of materials, greater flexibility in production and better quality products. However, as this is a labour-intensive industry, technological innovations are allowing for a reduction in its use, especially in segments that can be transformed into continuous processes, as is the case with the production of straight serial furniture (FERREIRA et al, 2008).

2.3 Safety and Health at Work

Occupational safety is defined by norms and laws. Occupational safety legislation is made up of regulatory standards, complementary laws such as ordinances and decrees, as well as the International Labour Organisation's International Conventions, ratified by Brazil.

An accident at work is one that occurs in the course of work in the service of the company, causing bodily injury or functional disturbance that can lead to death, permanent or temporary loss or reduction of capacity for work. They are equivalent to accidents at work (PANTALEÃO, 2015):

a. The accident that happens when you are providing services at the behest of the company;
b. Outside the workplace;
c. The accident that happens when you're travelling on company business;
d. An accident that occurs on the journey from home to work or from work to home;
e. Occupational disease (diseases caused by the type of work;
f. Occupational disease (diseases caused by working conditions;

Pantaleão (2015) states that accidents at work are mainly due to two causes:

Unsafe act: An act carried out by a person, usually aware of what they are doing, which goes against safety regulations. Examples of unsafe acts are: climbing on a roof without a safety harness, plugging in electrical appliances with wet hands and driving at high speeds.

Unsafe Condition: A condition in the work environment that poses a danger or risk to the worker. Examples of unsafe conditions are: electrical installations with bare wires, machines in a precarious state of maintenance, scaffolding on construction sites made from unsuitable materials

The term damage can be used to define any and all adverse changes to human health (such as injuries, illnesses or even death), to people's or companies' assets, or to the environment. It thus corresponds to adverse impact. In other words, damage is the negative consequence if an accident occurs. BSI-OHSAS-18001 (apud VASCONCELOS, 2006) also defines risk as: "Combinations of the probability of occurrence and the consequence(s) of a given hazardous event". It can therefore be said that risk corresponds to the ratio of probability versus consequence of an event occurring.

All employees need to have a correct understanding of the meaning of risk so that it can be identified and assessed appropriately. Barbosa Filho (2001 apud VASCONCELOS, 2006) shows that individuals' awareness and ability to recognise the possibilities of accident risk provide the minimum conditions necessary for them to actively collaborate in managing the environment in which they work.

Another important concept for Occupational Safety and Health (OSH) is that of an accident. There are various definitions of accidents, but the legal definition given by Law 8.213 of 24 July 1991 is: "anything that occurs in the course of work at the service of the company, or even in the course of work of specially insured persons, causing bodily injury or functional disturbance that causes death, loss or reduction of capacity for permanent or temporary work". This type of definition, in turn, only takes into account human losses. On the other hand, accidents don't always lead to personal injury.

Therefore, it is advisable to associate the concept of an accident with other damages, relating to materials, equipment, downtime, the time needed to re-plan activities, various social upheavals and other factors that hinder production and generate losses for the company.

Also related to accidents at work is the concept of an incident, which can be defined as the abnormal occurrence of a dangerous or undesirable event, but which does not result in damage, either due to control measures or other factors (CARDELLA, 1999 apud VASCONCELOS, 2006).

According to Vasconcelos (2006), occupational health and safety is understood as a set of technical and scientific knowledge aimed at preventing the occurrence of accidents and illnesses arising from the exercise of professional activities, and mitigating their consequences, protecting the integrity and working capacity of workers. In a broader sense, health and safety at work concerns not only employees, but also temporary workers, contract staff, visitors or anyone else who is in the workplace.

2.3.1 Ergonomics

In August 2000, the IEA (International Ergonomics Association) adopted the official definition, as ergonomics or Human Factors is a scientific discipline related to understanding the interactions between human beings and other elements or systems, and applying theories, principles, data and methods to projects in order to improve human well-being and overall system performance (VENTUROLI, 2002).

a. Physical ergonomics: This relates to the characteristics of human anatomy, anthropometry, physiology and biomechanics as they relate to physical activity.
b. Cognitive ergonomics: Refers to mental processes such as perception, memory, reasoning and motor response as they affect interactions between human beings and other elements of a system.
c. Organisational ergonomics: This concerns the optimisation of socio-technical systems, including their organisational, political and process structures.

2.3.2 Occupational health and safety regulations (NR's)

According to Pereira (2010), due to the various transformations that work has undergone, its form and its concept, studies have been developed that seek to explain the cause of workers' illnesses and their well-being in the workplace. Specific occupational health and safety professionals began to study and prevent accidents, illnesses and other factors that compromised the health and quality of the activities they carried out.

Concern about the well-being, health and safety of carpenters at work has been growing over the last few years, because when work is just an obligation or a necessity, the unfavourable situation increases greatly, posing risks to the physical and psychological integrity of carpenters and their team.

According to Venturoli (2002), there is a high risk of accidents, which can lead to the joiner being off work for considerable periods of time, which, as well as harming the employee, means losses for the joinery because, most of the time, there is no trained labour force to replace him, thus interfering with production and consequently the delivery of the furniture.

2.3.3 Technical standards for joinery activities

The technical standards that regulate procedures, materials and equipment in woodworking and related activities, as well as occupational safety, are governed in Brazil by the Brazilian Association of Technical Standards (ABNT), by the Fire Brigade of each state in cases related to fire and safety, by IBAMA and state and municipal legislation on environmental issues relating to the use of wood and the disposal of solid, liquid and gaseous waste:

    a. NBR-14048 - Furniture fittings and accessories handles and mirrors and key guides;

    b. NBR -15164 - Upholstered furniture sofas;

    c. NBR - 14033 - Kitchen furniture.

5 ANALYSING RESULTS

5.1 Company history

Bianchini (2014), the Bianchini family has a tradition in the timber and furniture business, which began in the 1960s in Urussanga/SC when Lauro Ângelo Bianchini opened his first joinery shop with his two brothers. In 1962 Lauro emigrated to Medianeira/PR, including a sawmill in his business. In 1963 he started a family and moved again, this time to São Miguel do Iguaçu/PR.

In 1983 he moved to the municipality of Cacoal/RO with his family and started Nelaze Ind. Com. Madeiras, in the current district of Riozinho. The 90s. In 1992, Móveis Bianchini Ltda was born, with his son Roberto Fernando Bianchini at the head of the business, working with solid wood, lacquered wood, sarrafeados, evolving to formica and MDF.

Over the years, Moveis Bianchini Ltda has sought to combine its know-how and tradition in the industry with the mechanisation and computerisation of its joinery, acquiring equipment and machines to make its products easier and more personalised. For company owner Roberto Bianchini, one of the main reasons he invested in automation was to increase the precision of cuts, speed up the production flow and increase work safety for workers in the furniture manufacturing process.

In 2012, the process of automating parts of the company's production sectors began, with the purchase of cutting machines, edge banding machines and software for drawing up projects. The process of automating the company took approximately four years, from the time the owners made the decision until the machines were fully operational.

According to the owner, there were no major difficulties in adapting to the use of the machines, but what caused the worker to be more resistant was the new trend in planned furniture, which was no longer produced with solid wood but with MDF, i.e. the abrupt change caused a certain amount of rejection on the part of the employee. The owner's plan is to automate the entire company within the next five years, as he believes that the investment in automation pays off when compared to the reduction in the number of employees and the speed with which the furniture is manufactured.

In order to stay abreast of market trends, the managers take part in trade fairs such as Casa Cor, held in various Brazilian states, and have a close partnership with local architects.

The planned furniture production process is characterised as a push system, meaning that an item is only manufactured once the order has been confirmed in the backlog.

For better visualisation and understanding, a flowchart of the production process was drawn up, as shown in figure 1.

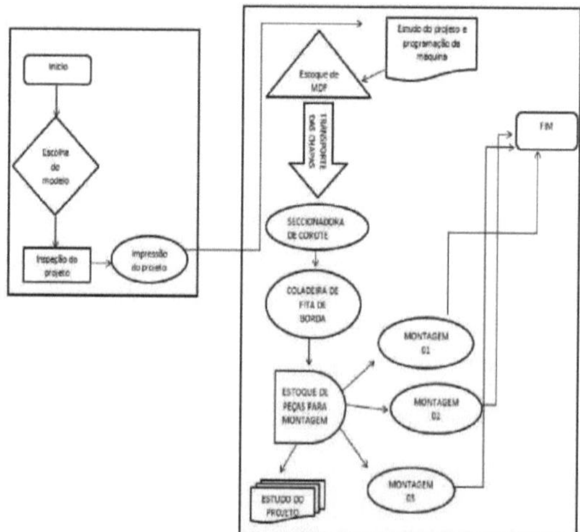

Figure 1: Production Process Flowchart

Source: Authors, 2016.

The production process at Bianchini planned furniture begins with a visit to the place where the customer wants to install the new furniture, taking actual measurements of the room. Then the virtual project is developed, which is idealised by the client using the Promob programme, carried out by the designer responsible for the order.

The designer then presents the virtual project to the client to check that everything is in line with the client's wishes and needs, after which an order is released for the purchase of the materials that will be used to make the furniture.

To continue the process, the specification of the cuts to be made by the machines is made, at which stage the right cut programme is used to assist the process. The project specifications are printed out and sent to the joiner, who studies the project and then separates the MDF formats that will be used.

At the cutting saw, furniture is produced from raw or coated chipboard panels. At this stage, the panels are sectioned according to the product's manufacturing process. Next comes edge banding, where the edges are part of the furniture's finishing process, covering exposed parts of the chipboard panel. After gluing the edge strips, the pieces are sent to the raw materials stockroom for processing. Only after the project has been analysed are the pieces sent to the assembly cells. In the final stage, the furniture is assembled at the customer's home.

5.2 Analysing interviews and questionnaires

The Bianchini furniture company has seventeen employees divided between the production, assembly, administrative and design sectors. The owner of the company also holds the position of manager, so an interview with fifteen open questions was applied to him. From analysing the owner's answers it is possible to conclude that before automation there were cuts and injuries to workers' limbs, with the fingers being the main areas affected, these incidents led to absences from work but never led to lawsuits for compensation. In order to gain a better understanding of the advantages and disadvantages of automation, questionnaires were administered to employees.

Following the proposed approach, questionnaires were only administered to employees who worked only in the company's automated sectors, noting that the company is not yet fully automated. The questionnaire was answered by five male employees aged between 19 and 49, where the employee who has been with the company the least has been employed for two years and the longest has worked in the same job for six years. Only one of the five interviewees worked in the company before automation, which explains why they unanimously reported that they had never suffered any accidents at work in the company before the machines were installed, because when they were hired the sectors they were employed in were already automated, and the longest-serving employee had never suffered any accidents either before or after automation.

Only one accident at work in the company after the machines were installed was reported in the survey. In an informal questioning, the machine operator said that one day, due to carelessness on his part, he started to press on his index finger, but the machine's emergency shutdown system was activated immediately and this prevented anything worse.

When asked about the disadvantages of using machines in the production process, the majority of workers said that there had been an increase in the repetitiveness of movements when carrying out tasks. This is a common occurrence in companies that have automation systems in their processes, as this greater repetition is due precisely to the agility of machines when carrying out tasks. This has been observed since the industrial revolution.

Still on the subject of the disadvantages of using machines, two operators reported an increase in physical effort, but after an in-depth analysis with the company manager, it was concluded that this was due to the fact that the company does not yet have skid-steer systems that move the MDF sheets around the factory floor, meaning that the operator has to pick up the sheet by hand and carry it to the machine where it will be processed, According to the company manager, this problem will soon be solved, as the company's new industrial facilities project will include a system of moulding machines that will carry out this work for the workers.

As for the advantages of using the machines, four of the five workers reported that they felt safe when carrying out their tasks on the automated machines. This safety is due to the sensing systems on the cutting machine and the emergency switch-off button on the edge banding machine.

6 FINAL CONSIDERATIONS

From the data collected and the analyses carried out, it can be concluded that there have been benefits for the workers and the company following the automation of the cutting, edge banding and software sectors, noting that the software sector has not been emphasised in this work. The implementation of the usikraft brand cutting machine and the seven brand edge banding machine has resulted in less time spent carrying out activities and greater safety for the worker when carrying out tasks.

The presence of automation has provided greater safety for employees, minimising the occurrence of accidents at work, but there has been an increase in repetitive movements due to the speed with which the machines are processed. The company manager says he is satisfied with the benefits brought about by the automation of his production process because three factors can be identified as a positive consequence of his investment: improvement in the quality of the final product because as the machines are less susceptible to error this has led to perfect fittings of the component parts of the furniture, no occurrence of worker absences due to physical problems caused by the performance of daily duties, and an increase in the company's productivity.

7 REFERENCES

ABIMAQ. **The History of Machines.** São Paulo, 2006.

ANTUNES, J. A. J. Foreword in - **Toyota Production System - more than just just-in-time.** Caxias do Sul. Educs, 1996.

BALLESTERO-ALVAREZ, M. E. **Organisation, systems and methods.** São Paulo: Mcgraw Hill, 1990.

BARBOSA FILHO, A.N.Segurança do Trabalho e Gestão Ambiental. 1ST EDITION. São Paulo: Atlas, 2001. p.17.

Bianchini planned furniture - Institutional. Cacoal, 2014 available at: <http://www.moveisbianchini.com.br/institucional>. Accessed on 08/06/2016.

BLACK, J.T. The Future Factory Project. Porto Alegre: Bookman, 1998.

CARDELLA, B. Safety at Work and Accident Prevention: a Holistic Approach: Safety Integrated into the Organisational Mission with Productivity and Quality. 1ST EDITION. São Paulo: Atlas, 1999. p.235

CARVALHO, Maria do Socorro M. V. de ; SOUZA, Gleim Dias de; LIBOREIRO, Manuel Alejandro Martínez. Supply chain management integrated with information technology   .
2006.   Available at   :
<http://www.scielo.br/scielo.php?script=sci_arttext&pid=S0034- 76122006000400010>.

CHIAVENATO, I. **Introdução à Teoria Geral da Administração.** 3 ed. São Paulo:

FERREIRA, M. J. B.,et al. **Furniture industry sector monitoring report volume I.** Campinas - SP, 2008.

GOMES, D. de O. GUIZZE, C. L.C. **Ergonomics in a small furniture factory: benefits for the company and workers.** XXXV Encontro nacional de engenharia de produção. Fortaleza - CE, 2015.

GROOVER, M. P. - **Automation Production Systems and Computer- Integrated**

Manufacturing. New Jersey. Prentice - Hall, 2001.

GUTIERREZ, R. M. V.; PAN, S. S. K. **Electronic complex: automation of industrial control** . Available at :< http://www.bndes.gov.br/SiteBNDES/export/sites/default/bndes_pt/Galerias/ArquivoA /conhecimento/bnset/set2807.pdf>. Accessed on: 15 January 2016.

LUZ, G. B.; KUIAWINSKI, D. L. **Mechanisation, Autonomy and Automation - A Conceptual and Critical Review.** XIII SIMPEP Bauru, SP, 2006.

MARTINS, P. G., LAUGENI, F. P. **Administração da Produção**. 1ª ed. São Paulo: Saraiva, 2002.

MOREIRA, D. A. **Production and Operations Management.** 5 ed. São Paulo: Pioneira, 2000.

MOREIRA, D. A. **Production and operations management.** São Paulo: Pioneira Thompson Learning, 2002.

OLIVEIRA. LUIZ, JOSÉ. **The importance of automation in industry and productive sectors** Available at: <http://www.ngeletrica.com.br/> 2015.

PANTALEÃO, Sérgio Ferreira. **Situations assimilated to accidents at work.** 2015. Available at <http://www.guiatrabalhista.com.br/tematicas/situacoes_acidentetrabalho.htm>

PAZOS, F. **Systems automation & robotics.** Rio de Janeiro: Axcel Books, 2002.

PEREIRA, P. F. **The Importance of Prevention in Occupational** Health and Safety - Occupational Health and Safety Technologist Feira de Santana - BA 09/2010. Available at: <http://www.ebah.com.br/content/ABAAABpV0AE/a- importancia-prevencao-na-seguranca-saude-trabalho>. Accessed on: 26 March 2013.

EDUCATION PORTAL. **Learn about the advantages of industrial automation.** Available at : <<http://www.portaleducacao.com.br/informatica/artigos/53605/conheca-as- advantages-and-disadvantages-of-industrial-automation> Accessed on: 15 June 2016.

SANTOS, J. J. H. **Industrial Automation.** São Paulo, 1979.

SILVEIRA, C. B. - **What is industrial automation.** São Paulo, SP, 2011. Available at:

<http://www.citisystems.com.br/o-que-e-automacao-industrial/>. Accessed on 13 May 2016.

SLACK, N.; CHAMBERS, S.; JOHNSTON, R. **Administração da Produção.** 2 ed. São Paulo: Atlas, 2002.

TIGRE, P.B. **Gestão da Inovação: A Economia da Tecnologia no Brasil.** Rio de Janeiro: Elsevier Publishing, 2006.

TUBINO, D. F. **Manual de Planejamento e Controle da Produção.** São Paulo: Atlas, 1997.

VASCONCELOS, F. D. L., et al. **Safety Management in Construction - Discussion on a System Model. Proceedings.** CMATIC, Recife - 2006.

VENTUROLI, F. **Ergonomic analysis of the work environment in joinery shops in the Federal District.** 55p. Dissertation (Master's in Forestry Sciences) - University of Brasília, Brasília, 2002.

CHAPTER 6

POSITIVE ASPECTS OF AUTOMATION FOR PRODUCTION IN A PLANNED FURNITURE FACTORY IN THE MUNICIPALITY OF CACOAL/RO

André Jun Miki
Priscilla Lidia Salierno
Ezequiel José Hottes
Ismael Josué Hottes
Juander Antônio de Oliveira

SUMMARY

This article presents production system concepts, with a focus on customised production systems. It also provides an insight into automation and its main advantages. The production system is made up of a sequence of activities for the manufacture of a good or service. Its efficiency and effectiveness are being optimised with the help of automation, which brings benefits such as: agility, cost reduction, safety, improved quality, standardisation, among others. The study aims to analyse the advantages of automation for production at the Bianchini Móveis Planejados company in the municipality of Cacoal - RO. According to the article's general objective, this research is classified as descriptive, with a qualitative approach and deductive methods. In the first stage, a bibliographical survey was carried out in books and articles. The second stage involved a semi-structured interview with the manager of an organisation in the furniture sector. In the third stage, the results were analysed in accordance with the specific objectives using verbal extracts collected during the interview. The research follows ethical aspects, thus guaranteeing the integrity of the information. The research site is in the municipality of Cacoal/RO, where the company Bianchini Móveis Planejados is located. The results obtained were that automation is essential for the company to remain in the market, as it is extremely competitive. The main benefits of its application in the production process at Bianchini Móveis Planejados are, according to the manager/owner, improved quality and productivity, as well as making it possible for employees to work in the factory, since before the implementation of automation they only worked on commission.

Keywords: Production system, Automation, Advantages.

1 INTRODUCTION

The advance of technology in the world has brought countless advantages to human beings, but it has also made the market increasingly competitive. For companies to stay on the market, they need to keep up to date with technology. Automation is an example of this advance, and is indispensable for those who want to increase their efficiency and productive effectiveness, as it brings benefits such as: improved quality, reduced costs, safety and competitive advantages.

Automation is made up of the set of tools needed to produce a particular item, with the aim of handling and controlling the process, so that there is little or no human intervention in the respective stage of carrying out the work (SILVEIRA, 2011). But to understand automation, you need to know a little about production systems, which are "a set of interrelated elements (human, physical and management procedures) that are designed to generate end products whose value exceeds the total costs incurred to obtain them" (FERNANDES, 2010, p. 01).

The aim of this work is to analyse the advantages of automation for production at Bianchini Móveis Planejados. To do this, it is important to identify the company's production system, as well as describing the benefits of using automation that have been identified for companies that have decided to automate their production process.

Therefore, the relevance of the study lies in demonstrating, through an interview with the owner, how important automation is for a Bianchini Móveis Planejados company, so that its growth prospects can be achieved.

2 THEORETICAL FRAMEWORK

2.1 Basic systems concepts

Antunes (2008) states that the word "system" is an abstract form used of a relatively complex situation involving physical, chemical and biological elements. In this way, Antunes (2008) defines a system as a group of parts that operate together to achieve a common objective for all, i.e. a group of related components that work together to achieve a goal, in a process of transformation.

Contributing to the conceptualisation, Chiavenato (2003) states that a system is the idea of a set of elements interconnected to form a whole. In this way, it is understood that the whole has its own characteristics that cannot be found in any single element. The author uses as an example the characteristics of water, which in turn has completely different characteristics when compared to the separate elements, in this case oxygen and hydrogen.

Chiavenato (2003) also provides other definitions for systems, such as:

A) A system is a set of elements in reciprocal interaction;

B) A system is a set of joined parts that relate to each other to form a whole;

C) A system is a set of interdependent elements whose final result is greater than the sum of the results that these elements would have if they operated in isolation;

D) A system is a set of interdependent and interacting elements aimed at achieving a goal or purpose;

E) A system is a group of units combined to form an organised whole whose characteristics are different from those of the units;

F) A system is an organised or complex whole; a set or combination of things or parts, forming a complex or unitary whole oriented towards a purpose.

2.2 Production System

A production system is "a set of interrelated elements (human, physical, and managerial procedures) that are designed to generate end products whose value exceeds the total costs incurred to obtain them" (FERNANDES, 2010, p. 01). This means that the production system is made up of a sequence of activities for the manufacture of a good or service.

In the process of transforming raw materials into a final product, it is common to use the term input and output, which means input of inputs and output of the final product. In a production system, the output can be either the production of a good or the provision of services. Therefore, when the output of a production system is a good, it is called a manufacturing system, while if it is a service, it is called a service system (FERNADES, 2010). Figure 1 shows this process:

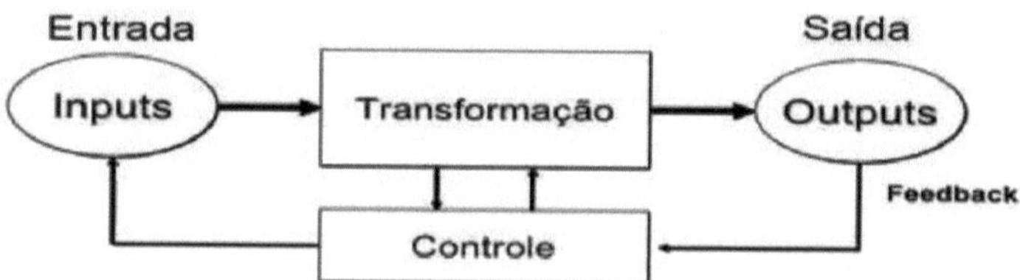

Figure 1: Simplified representation of a production system

For Shingo (1996), the evolution of the production system is based on the principle of five industrial revolutions:

A) First revolution: progress through the division of labour;

B) Second revolution: increased hand function (mechanisation and motorisation);

C) The third revolution: the science of labour;

D) Fourth revolution: responding to human needs (Hawthorne method);

E) Fifth revolution: development of the zero stock era.

Fernandes (2010) also emphasises that a production system is considered effective when the objective or goal is achieved. And efficient when

resources are utilised in the best possible way, i.e. without waste. However, it is necessary for the company to have a performance meter, because it is through the meters that the company will be informed whether the production system is efficient or effective.

However, it's impossible to talk about production systems without mentioning the three main types of production system that have been applied in industrial production chains (automobiles): Taylorism, Fordism and Toyotism (PENA, 2016):

A.   Taylorism: developed by Frederick W. Taylor (1856-1915), its main function is maximum productivity through repetitive patterns of workers and machines, as well as the division of tasks to optimise work, known as the mass production system;

B.   Fordism: developed by Henry Ford (1863-1947), it had practically the same thinking as Taylor, however, it had some differences. One of the main differences was the use of conveyor belts in the production chain, which allowed for greater agility and productivity;

C.   Taylorism: developed by Taiichi Ohno (1912) and Eiji Toyoda (1913-2013), in order to overcome the crisis in Japan and remain on the market, Japanese companies had to remodelling a production system. This type of system is the opposite of Fordism and Taylorism, as it is based on the following premises: flexible production that varies according to demand, multifunctionality of workers, zero stock and no standardisation of products.

Studying and understanding the production system is extremely important, because in this way it is possible to determine which direction to take in designing or determining or restructuring production systems, since the changes brought about by constant innovation cannot be ignored (SHINGO, 1996). The following topic will explain one of the production system classifications, the intermittent production system.

2.3 Discrete or Intermittent Processes

The word intermittent means a process in which there are interruptions, which ceases or restarts at intervals. With this in mind, the discrete or intermittent production process is made in batches, i.e. once a batch of a product is finished, other products take its place on the machines, so the original product won't be made again for some time. An example is the small-scale manufacture of textiles and metallurgical plants, which divide operations into stages and on the same machine (CAMPBELL; BERÇOT; TEIXEIRA, 2011). As for the provision of services, we can mention restaurants, which also work with an intermittent system.

According to Cauzzo and Bianchi (2005, p. 256), "the intermittent batch production system is characterised by producing the same product several times, usually in specified batch sizes, and when production ends, other products take its place for the production of another batch". This type of production occurs when there are several types of product, which are quite different, and which require a frequent change in the product to be processed by each processing centre (GONÇALVES, 2004). However, there is another classification for the intermittent system, which is usually called an intermittent on-demand system, which will be the subject of the next topic.

2.3.1 Customised intermittent production system

In a highly competitive environment, companies need to invest in innovation and face up to today's challenges. Among these challenges is the elimination of stock so that overproduction does not occur. Shingo (1996) states that some managers consider stocks to be a "necessary evil", but this is a big mistake, as they can lead to major losses.

Overproduction is avoided when production is carried out at the right time and on time. This philosophy is known as Just in time, which is nothing more than producing without excess or meeting demand instantly. Therefore, demand forecasting measures must be accurate if production scheduling is to be efficient. It then suggests that products should be manufactured according to demand, as well as being of optimum quality (SHINGO, 1996).

The term "made to order" is often used to mean various things, meaning that there is no precise definition. However, this type of system is geared towards meeting the specific needs of customers, and is characterised by being a process that has low demand and high variety (TUBINO, 2009).

According to Silveira (2012), a made-to-order system is a process in which the customer dictates the pace of production, and for professionals in the field it is considered a pull production system. Silveira (2012) also reports that the biggest advantage of this system is the reduction in losses and the possibility of customising the customer's product. However, production planning is more complex due to the allocation of the various resources available, as a customised product requires the work of several professionals, since it is necessary to ensure the product's delivery date.

This type of system requires highly flexible production resources focused on meeting customer specifications. In this way, certain idleness occurs in production, due to the waiting time in which the product is demanded by its consumers, thus generating an increase in the final cost of the goods or services. Examples of this system in the production of goods are the manufacture of ships, aeroplanes and hydroelectric power stations, among others. As for the provision of services, there are advertising agencies, law firms, architects, etc (TUBINO, 2009).

2.2 Automation

According to Goeking (2010), automation emerged as a way of reducing the involvement of the "human hand" in industrial processes. In the 10th century, the hydraulic mill was used on a large scale to supply flour. It was one of the first human creations to automate work, and this process of mechanisation was what drove the emergence of automation. As can be seen in the image below.

Figure 2: Hydraulic flour mill

Source: Goeking, 2010.

According to Goeking (2010), the mill led to an increase in food production that had never been seen before. The mill had the capacity to replace 10 to 20 men. As a result, man has sought knowledge in technology to develop his manual labour activities. One of the first examples was the creation of the steam engine, which was used to power industrial equipment. In 1775, the need to produce culminated in the industrial revolution, which was a major development in replacing manual labour with machines and this accelerated the development process in the production sector.

For Silveira (2011), the concept of automation is the combination of the tools needed to produce a particular item, with the aim of handling and controlling the process, so that there is little or no human intervention in this work. Currently, the word industrial automation refers to any type of advanced mechanisation or synonym for technological evolution, which is related to the cyber world, where technological advances are constantly evolving.

According to Lamb (2015), automation is the use of programmable logic commands and mechanised equipment to replace activities that used to involve human commands. Automation goes beyond mechanisation, as it reduces the need for human sensory and mental requirements, as well as optimising productivity. In the 1940s, the term automation was coined by a engineer at Ford Motor Company, where he described various systems that could replace human effort and intelligence.

After a brief conceptualisation of automation, the following text will explain the application of this type of technology in the industrial field and how these automated mechanisms help to increase production and reduce costs in industrial processes.

2.2.1 Industrial Automation

According to Venturelli (2014), industrial automation is a set of techniques and technologies that are applied to the industrial area, using physical control equipment (hardware) and logical command and control programmes (software), with the aim of making processes automatic, seeking production on a scale that is stable and promotes safety in the operational sector.

There are several pieces of equipment used in automation, and for Parede et al. (2011) the programmable logic controller (PLC) is one of the most important. Around the 1960s, the PLC emerged and revolutionised industrial controls. In that decade, automation was still in its infancy and there was an extensive amount of wiring interconnected by circuits that were run entirely by relays with a fixed logic base. Dick Morley, an employee of Bedford Associates, created the first PLC in 1968 to replace the cabinets used to control sequential and repetitive operations.

According to Parede et al. (2011), in the 1960s wiring and relays meant that automation had some disadvantages, such as:

A) **Inflexibility:** because changing the production process meant changing the entire control panel, which would mean changing all the logic, and this was very time-consuming.

B) **High operating costs**: the panels were very large and therefore required a lot of floor space to use them.

C) **High development and maintenance costs:** the logic of the relays had to be minimised, as this would reduce the cost involved in assembly.

Fernandes et al (2002) points out that companies with small and medium-sized sales have between 75% and 85% interest in automation, while companies with large sales have 33% interest.

Fernandes et al.'s (2002) view is not erroneous for the time; today we can see that interest in automation is global, due to the great advances in technology and with this we can see the benefits of using programmable control with logic, such as the reduction in production costs and labour is one of the advantages of using automation. Ladeia (2014) says that the expansion of digitalisation and automation would reduce the number of workers in factories. So automation has its good side in the factory sector, but in the economic sector in general these days it can have some disadvantages.

Silva (2007) tells us that automated systems can be applied to simple machines or industries. The difference lies in the number of elements monitored and controlled, known as "points". So automation is present in small electronics that the human being doesn't notice, such as an "electric oven", where the command is given, but all the logical functions for that command have already been programmed. A simple automation diagram is shown below.

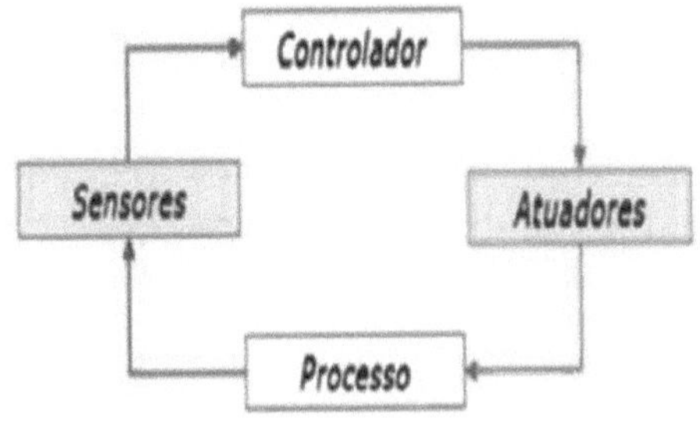

Figure 3: Simple automation diagram

SOURCE: **HTTPS://**CLPREDES.FILES.WORDPRESS.COM/2010/05/CLP-FIG-1.JPG

Silva (2007) explains the process of the simple diagram, where the "sensors" are the elements that provide information about the system and the "actuators" are responsible for carrying out the work in the process to which automation is being applied, the "controller" is the element responsible for activating the actuators, taking into account the instructions passed on by the sensors and in order to have complete automation it is necessary to study these four items.

2.2.2 The benefits of automation

Silveira (2014) states that there are various types of industrial automation systems, but not all the types of automated systems needed in a factory are available on the market. In order to automate, industries must first design, evaluate and purchase, as all kinds of needs must be put on the table, so that there is great suitability for different demands.

However, Silveira (2014) presents seven benefits of implementing an automated system, such as:

A) **Increased productivity:** automation applied to automatic machines makes it possible to achieve faster production cycles with greater efficiency and repeatability. So a person can't do the same movement all day with the same perfection, but with an automated system, it can be done with

71

the same precision.

B) **Cost reduction**: it is able to reduce installation costs, as it can bring a return on investment through increased productivity. With an automated system, human labour will no longer be needed, as the system is designed to carry out different production methods without the need for human hands.

C) **Improved quality**: when you have automated machines, the result is more consistent and repeatable, in turn eliminating problems caused by human error, and processes are carefully regulated and controlled, making the quality of the product more consistent.

D) Safety: an automated system project has to be based on the premise of safety, because organisations are designed to reduce accidents, which obliges manufacturers to follow safety practices. As the system is operated by a computer, the chances of accidents are very low, so automated machines are able to work in extremely hot environments, thereby reducing the risks to human health.

E) **Competitive advantage**: with industrial automation, companies are now in line with or even ahead of their competitors, so automation enables companies to remain stronger in the face of economic turbulence and threats from competitors.

F) **Accuracy**: as the processes are measured on the main computer, which has an artificial intelligence programme embedded in it, it can be seen that without this type of programme it would be impossible to maintain good accuracy, so when the system is in use, all kinds of sensors and processors are used to monitor the entire process and thus have good accuracy in the tasks.

G) **Remote monitoring**: the systems allow an operator to control production processes from a certain distance. With an internet connection, it is possible to communicate over a longer distance.

The benefits offered by industrial automation are very essential in the manufacturing sector, so acquiring this automated equipment would increase production and, consequently, profit for the entrepreneur. Given that the market is modernising, adapting technology will be one of the strong points for staying in this modernised market.

3 METHODOLOGY

According to the article's general objective, this research is classified as descriptive, with a qualitative approach and deductive methods. The first stage involved a bibliographical survey of books and articles. In the second stage, a semi-structured interview was conducted with the manager of an organisation in the furniture sector. In the third stage, the results were analysed in accordance with the specific objectives, using extracts of verbalisation collected during the interview. The research follows ethical aspects, thus guaranteeing the integrity of the information.

4 RESULTS AND ANALYSIS

Based on the concepts acquired above about automation, a survey was carried out in the planned furniture sector, where its production system was analysed and the benefits of automation for production were subsequently identified. An interview was therefore conducted with the owner/manager.

4.1 A brief history of the company

The company where the research was carried out is known as Bianchini Móveis Planejados and is located at Rua Machado de Assis, 1822 in the Industrial neighbourhood in the city of Cacoal - RO. As its name suggests, it is a company that manufactures customised furniture, bringing innovation, quality and sophistication to its customers, which ensures that the product stands out from ordinary furniture.

The administrative manager, who is also the owner, is called Roberto Fernando Bianchini. He is 46 years old, has a degree in administration and is an accountancy technician. Bianchini has been operating in the customised furniture market for 22 years and is considered one of the pioneering companies in the municipality of Cacoal. Today it has a total of 17 employees.

However, in order to become a company as it is today, it went through a series of adaptations due to changes in the market. At first, Roberto Fernando Bianchini's father had a joinery business, but changes in legislation made it difficult for him to stay in the business.

After getting to know MDF, Roberto decided to work with this material so that he could make furniture the way his customers wanted it, i.e. planned furniture. From then on, Bianchini furniture was created.

4.2 The organisation's production system

As mentioned at the beginning of this article, the make-to-order system is when the customer dictates the pace of production. It is also characterised by producing items out of series, with a wide variety in its production mix. In view of this, it is possible to classify the production systems of the Bianchini planned furniture company as a made-to-order production system, so that its product can be differentiated and of excellent quality, thus meeting customer expectations. However, these products are considerably expensive, which makes them exclusive to middle and upper class customers. Below are the production processes of the Bianchini planned furniture company.

4.2.1 Production process

The raw material used in the production process of the planned furniture is Medium Density Fiberboard (MDF). It is purchased in southern Brazil and in the municipality of Cuiabá (MT). The thicknesses purchased are 15 mm and 20 mm and are stored within the production process. The owner/manager also says that when demand is high and the MDF from the south doesn't arrive in time, the purchase is made in Ji-Paraná (RO), because production can't be interrupted due to a lack of raw materials.

Production is done at the customer's pace, so the product is modelled according to what is specified in the way the customer wants. The designs are made in two software programmes called: Promob Kot and Corte certo, the 3D drawing is processed, endeavouring to meet all the customer's requirements. Production only begins once the project has been finalised.

Production begins with the purchase of the raw materials, then they are stored in stock in the factory, the material is cut by the sectioning machine as stipulated by the project, once the cut has been made, edge tape is applied to the side of the MDF to finish it off and finally the assembly and delivery to the customer's home is carried out. Figure 4 shows the stages of the company's activities.

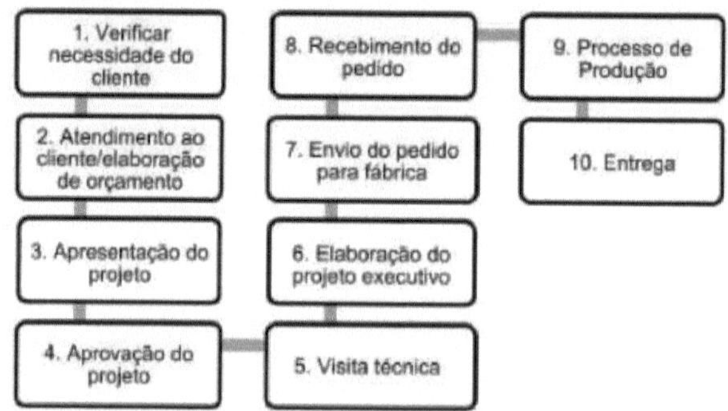

Figure 4: Flowchart of activities for organising the custom-made furniture process

Figure 4 shows the flowchart of the most extensive activities in the production process at Bianchini Móveis Planejados, making it easy to visualise the company's entire product manufacturing process. The first procedure (figure 4) is to check the client's needs, followed by drawing up a budget. After presentation and approval, the project is sent to the factory, but before manufacturing begins, a technical visit is made to the client's home to check the feasibility of the project. Finally, the final product is manufactured and delivered to the client.

4.3 Automation in Production

The automated machine is the sectioning machine (figure 5) and all the cutting of the part is done according to what was stipulated in the project.

Figure 5: Cut-off switch machine

The cost of installing this type of machine is high, as an automated process is still expensive and requires specialised labour. The machine brings great benefits to production, as it increases the company's production capacity, making a complex job easier.

4.3.1 Analysing the interview

The literature shows that automation brings numerous benefits, such as increased productivity, cost reduction, improved quality, safety and precision, among others. It is therefore necessary to assess how automation has benefited the customised production system, taking into account the Bianchini planned furniture company.

According to manager/owner Roberto Fernando Bianchini, automation has made it possible to employ staff, who were previously only paid per unit produced. However, the number of employees has been reduced, and before applying automation, three times the number of employees were needed today to do the same job.

The owner/manager also points out that one employee was involved from start to finish in the production of a product, i.e. he was multifunctional. They worked on commission and when they got paid, they started to miss a lot, which greatly jeopardised production and meeting demand. However, with automation it was possible to reverse this situation, bringing the employee back to the company after signing their cards, thus receiving their pay on a monthly basis. In this way, it is necessary to train employees to carry out their work. Another point to note is that employees are not now involved in every stage of production, making them functional. Roberto also emphasises that the workers can't "think", they just have to execute and that the management is responsible for this.

To summarise, according to the owner/manager, the use of automated systems has brought major improvements in production for the Bianchini company, especially in terms of quality, precision and productivity. He also emphasises that automation is essential for those who want to stay in the market, given the fierce competition. And in order to stand out from other companies, it is essential to have a quality product, which can be acquired through automation.

Regarding satisfaction with automation, the owner/manager is favourable, but he intends to increase the level of automation in the company by investing in other automated machines. It's an automatic cutting machine that does the work of two employees, thus reducing his labour costs.

Automation is very important for the production of goods, but it is not without its critics. For the owner/manager, the main criticism of automation is that it takes the artisanal part out of the process, as well as having a high initial investment. But even so, it pays to invest, as it pays for itself in just a few years. Therefore, the owner/manager recommends automating the production system to any company.

5 CONCLUSION

Automation has been gaining market share and one of the reasons is the great advance in technology, in which entrepreneurs want to reduce costs in their companies. So many companies are looking to install this type of system in their production processes.

An automated production process provides entrepreneurs with better product quality, as it brings precision to the means of manufacturing a product.

product. Automation has been very favourable for the Bianchini company where, according to owner Roberto Fernando, it has increased production and reduced labour and raw material costs.

Another factor to be addressed is the training of employees to carry out their activities. Roberto Fernando used to let each person carry out various activities in the company and he saw that this wasn't being favourable and adopted the idea that each employee had to have a role in each production process, so he decided to train each employee for each process, so it was seen that in order to automate a process, training is needed so that there are no errors in production.

So having an automated process is expensive, this study points out that for the company, the advantages of applying it to the production system are very great: increased productivity, reduced costs, improved quality and safety. In this way, the company can stand out from other companies in terms of quality, productivity and safety. This shows that an automated process makes production more efficient and in turn increases production.

5 REFERENCES

ANTUNES, J. et al. **PRODUCTION SYSTEM - Concepts and practices for lean production design and management**. Bookmam, Porto Alegre, 2008.

CAMPBELL, M.; BERÇOT, M.; TEIXEIRA, M. A. **PRODUCTION ADMINISTRATION - Production Systems**. Faculdades Integradas de Jacarepaguá (FIJ), 2010.

CAUZZO, J. S.; BIANCHI, R. C. **Sistema de Produção nas Organizações de Pequeno, Médio e Grande Porte no Sector Calçadista do Vale do Rio dos Sinos**. Applied Social Sciences, Santa Maria, v. 1, n. 1, 2005.

CHIAVENATO, I. **Introdução à teoria geral da administração**. Editora Elsevier, 7ª edition, Rio de Janeiro, 2003.

FERNANDES F. C. F. ET AL. **Industrial automation and computerised production management systems in market foundries**: Gestão e Produção, 2002.

FERNANDES, F. C. F. **Planejamento e controle da produção: dos fundamentos ao essencial.** Editora Atlas S. A. São Paulo, 2010.

GOEKING W. **From the steam engine to automation software**: Portal to the electricity sector, 2010. Available at:<http://www.osetoreletrico.com.br/web/a-revista/343- xxxx.html> accessed on 01/05/2016.

GONÇALVES, J. F. **SISTEMAS PRODUTIVOS - Gestão Da Produção**, 2004. Available at: <http://www.fep.up.pt/docentes/jfgoncal/PDF's/1 G302/Por/1 -INTRO. pdf> Accessed on: Apr/2016.

LADEIA B. **Industry is sexy again**: Brasil Econômico, 2014. Available at <http://economia.ig.com.br/empresas/2014-09-26/a-industria-voltou-a-ser- sexy.html> accessed on 01/05/2016.

LAMB F. **Industrial automation in practice: Axis and control of industrial processes**. Lookman, tekne series, 2015.

PAREDE I. M.; GOMES, L. E. L.; HORTA, E. **Eletrônica: Automação industrial**. Padre Anchieta

Foundation, V. 06, Paula Souza Centre, Government of São Paulo, 2011.

PENA, R. F. A. **Production system**. Mundo Educação, 2016. Available at < http://mundoeducacao.bol.uol.com.br/geografia/sistemas-producao.htm> Accessed on: 30/04/2016.

SHINGO, S. **The production system from the point of view of production engineering**. Bookmam, 2ª edition, Porto Alegre, 1996.

SILVA M. E. **Curso de Automação Industrial**, Piracicaba, 2007.

SILVEIRA C. B. **What is industrial automation**: Citisystems, 2011. Available at <http://www.citisystems.com.br/o-que-e-automacao-industrial/> accessed on 01/05/2016.

SILVEIRA C. B. **Seven benefits gained through Industrial Automation**: Citisytems, 2014. Available at < http://www.citisystems.com.br/sete-beneficios- automacao-industrial/> accessed on 01/05/2016.

SILVEIRA, C. B. **Sistema De Produção, Agregação de Valor e Automação**. Gestão da Produção, Citisystems, 2012. Available at: < http://www.citisystems .com.br/sistemas-producao-automacao-industrial/> Accessed on: Apr/2016.

TUBINO, D. F. **Planejamento e Controle da Produção - Teoria e prática**. Editora Atlas S. A. 2ª edition, São Paulo, 2009.

CHAPTER 7

AUTOMATION APPLIED TO STOCK MANAGEMENT IN A SUPERMARKET

Alessandro Aguilera Silva
André Jun Mik
Graziela Luiz Franco Martinez
Fabiana Maria Alves
Monique da Silva de Sousa

SUMMARY

Commercial automation has led to major transformations in stock management, especially in the retail trade (retail shops and supermarkets). The implementation of computerisation redefines commercial operations, the volume of stock and logistics flows, as well as leading to a growing rapprochement between industry and retail. With this in mind, the aim of this study is to identify the results obtained with the application of automation to stock management in supermarkets, to verify the improvements obtained with its implementation, what problems the supermarket faces after the implementation of automation and to analyse the results obtained with the application of automation to stock management in supermarket X. This is a descriptive, qualitative study using the deductive method, as its aim is to describe behaviour or define and classify facts and variables. Initially, bibliographical research was carried out in books and articles on the subject, followed by interviews with the application of a questionnaire with closed questions to the following departments: butchery, billing, cold cuts, produce and grocery. Interviews were carried out with those responsible for the Information Technology (IT) and Human Resources (HR) sectors, using a questionnaire with open and closed questions. From the results obtained, it was analysed that automation has contributed to better stock management so that purchases comply with the company's stock policies.

Keywords: Stock management, Commercial automation, Supermarket.

1 INTRODUCTION

Currently, the Brazilian automation market has brought about transformations that directly impact users such as the retail trade (retail shops and supermarkets), improving the efficiency of their operations. In order to increase sales and turnover, this sector must adopt new management and commercial strategies with the adoption of inventory management tools by implementing technology (BASTOS and SEGRE,1999).

The restructuring of the sector in order to gain a competitive advantage is taking place through the use of new stock management systems, such as the adoption of policies, tools and techniques like stock management. Computerisation is an increasingly important element in the retail chain and in supporting distribution activities. The implementation of computerisation redefines commercial operations, the volume of stocks and logistics flows, as well as leading to a growing link between industry and retail (SANTOS, 2002).

With the use of technological resources, companies can have new commercial opportunities, allowing them to expand into new markets or new segments of existing markets (BAZZOTTI and GARCIA, 2007).

According to Porter and Millar (1985), previously the use of computers in information processing was focused on accounting and its main function was to record it. But today, information processing can change the value chain, improve and control functions. The adoption of technology has become a tool for gaining mandatory competitive advantage in the face of commercial expansion, increased competitiveness, the popularisation of technology and easy access to information.

In this new, totally competitive environment, the retail sector has been trying to adapt to new trends, especially e-commerce, commercial automation, Electronic Data Interchange (EDI) and Efficient Consumer Response (ECR), all based on the use of Information Technology (IT). Technological trends are forcing retail managers to re-evaluate their businesses and adapt to this new scenario (BACHEGA and ANTONIALLI, 2005).

The problem of this research is based on the premise of commercial automation applied to stock management in a supermarket and the importance of real-time information when purchasing new items for stock, combined with the use of technology to facilitate the development of administrative activities, record information and change the value chain, perfect and control functions. Today's commerce can no longer support the use of the passbooks and mechanical cash registers that were used in the past. With competition between retailers and more demanding customers, the implementation of commercial automation in stock management has become a fundamental strategic tool in stock management, a solution for retailers who are constantly evolving, seeking to improve their business and improve profits and customer satisfaction.

2 THEORETICAL FRAMEWORK

The history of industrial automation began in the 18th century with the British invention of the steam engine, which increased the productivity of manufactured products. With the growth of industries in the following century, the replacement of energy sources and iron with steel fuelled the growth of European and American (USA) industries. In the years that followed, mechanical devices called relays were created that would soon take over factories. These events and others that followed have been called the Second Industrial Revolution (SILEVIRA; LIMA, 2003).

In the 20th century, factory environments still had very rudimentary automation processes. But in 1909, Henry Ford's idea changed the thinking of the industry of his time, which has continued to this day. Henry devised the assembly line, which may have been the beginning of industrial automation. From the application of his idea in industry, new production concepts emerged, for example: mass production, assembly points and intermediate stocks (SILEVIRA; LIMA, 2003).

By the middle of that century, General Motors (GM) was producing cars on an economy of scale, and in the years following Henry's death, GM already had automated machines based on relay logic, incorporating great attributes of technological innovations, requiring the installation of hundreds of electromechanical devices and power semiconductor devices mounted on electrical command and control panels and cabinets, providing great interconnectivity and efficiency in the use of electrical energy (SILEVIRA; LIMA, 2003).

In 1968 Dick Morley, an employee of Bedford Associates, created the first Programmable Logic Controller or Programmable Controller, also known by its acronym CLP or CP, with the aim of replacing the cabinets used to control sequential and repetitive operations on the assembly line of the General Motors car industry, replacing the heavy electromechanical relays. MODICON (Modular Digital Controller) was the first Programmable Logic Controller invented and replaced all the paraphernalia, making the system much more flexible, economical and efficient (SILEVIRA; LIMA, 2003).

2.1 Automation

The concept of automation spread with the industrial revolution with the construction of machines. It is linked to automatic, repetitive mechanical movement and is a set of techniques through which various systems are built for the efficiency of production activities, information management, stock management, among others. It is characterised by processes that have little or no interference from human labour (DUARTE, 2013).

The role of automation is to interact the information in the process in order to achieve excellence in the use of resources, by managing the information resulting from each task carried out and resizing or reorienting the next stage, with the aim of achieving the final result.

2.1.1 Commercial automation

Commercial automation is the replacement of manual processes with the application of tools to automate commercial processes, in other words, to make them easier. The management of commerce with the integration of man and machines in management, seeks to reduce labour, cost expenses, as well as more solid management (DUARTE, 2013).

According to Duarte (2013), the implementation of commercial automation helps to reduce errors such as: calculating and entering prices, quantities, writing a cheque, issuing an invoice, making these activities safer and more efficient, while at the same time improving the work of employees and customer service.

With the growing competitiveness of the supermarket segments, automation is not only a viable alternative for guaranteeing good prices for the product mix and differentiated service, it also helps to control stock and costs (VILELA, 2006).

In the supermarket sector, the implementation of information technology (here referred to as commercial automation) has been a strategy for guaranteeing the level of service offered to customers with a view to increasing competitiveness and operational excellence, facilitating the development of administrative activities, improving the way in which activities are operated that are perceived by customers, recording information and changing the value chain, perfecting and controlling functions (VILELA, 2006).

The implementation of Information and Communication Technology (IT) in supermarkets makes it easier to identify the profile, to offer more variety in the supermarket, to speed up interaction between suppliers and outsourced companies and to improve customer service (BASTOS; SEGRE 1999).

2.2 Stock Management

For the logistics system, stock management is of paramount importance, comprising various operations that have a significant impact on customer service levels and total costs (WANKE, 2011).

Retail-orientated inventory management means determining the optimum lot size for purchases in the light of different economic constraints. Stock decision-making can lead organisations to adopt stock policies to react to demand. And according to Fleury, Wanke and Figueiredo (2000), defining an inventory policy basically depends on four questions: how much to order; when to order; when to keep stocks and where to locate them.

According to Wanke (2011), stock policies play an important role in the flow of products, guaranteeing good service to the end consumer without excess or lack of products in stock. Stock policies are based on:

The policies and quantitative models used: react or anticipate - how much to order on the predefined dates, economic lot size, replacement point and replacement interval. According to the bargaining power of suppliers and customers, capital intensity;

The organisational issues involved: and the management model adopted, such as integrated planning systems, management models, organisational culture;

1. Type of technology used: level of automation;
2. Monitoring process performance.

Inventory management consists of determining when and how much, managing your processes according to demand in order to meet customer expectations, reducing costs and generating profit. While the uncertainties of demand force companies to maintain their safety stocks, reducing stock levels is a decision that requires caution. According to SUCUPIRA (2009), stock management consists of registering the items to be controlled, stock planning, control and planning feedback.

Registration of materials: identifying the items to be catalogued, standardising nomenclature to allow information on individual items to be consolidated, classifying and cataloguing by giving a code for processing through an electronic system that registers all the items to be controlled used by the company;

Stock planning: planning where you determine the dates quantity of incoming and outgoing products and the replacement point. Times when stock policies are defined for each product according to demand and delivery times, etc;

Stock control: based on data records and the movements and changes in stock situations;

Feedback: this consists of comparing the control data with the data established in the planning, in order to detect deviations and investigate possible causes. It is the role of stock managers to revise or correct the planning established in order to achieve the objective of stock control.

2.3 Just in Time

According to Ohno (1997), just in time corresponds to the right moment or opportune moment, but in English it means on time, not at the exact moment, but a little earlier so that there is some slack. It is one of the pillars for implementing lean manufacturing in companies. Production only takes place when the need arises to meet customer demand.

2.4 Kanban

According to Figueiredo (2010), kanban, of Japanese origin, means control by means of cards, signs or visual management, signalling the production or movement of an item as the main tool for implementing a pull production system, and is the main characteristic for implementing just in time (JIT).

At Toyota this card is a piece of paper used inside an envelope containing information divided into categories, such as collection, transfer and production information, with the main objective of controlling and reducing inventories in factories, being produced when the customer requests it. It is divided into three colour levels which vary according to the company and are generally identified as (green, yellow and red) identifying cycle stock (quantity, buffer stock and safety stock) (OHNO, 1997).

According to Tubino (1999), kanban began to be used in the 1960s by engineers at Toyota Motors with the aim of improving productivity throughout their system and involving the workforce. It can be conceptualised as a technique for managing materials and production, with each item being produced in the quantity needed and at the right time, by means of signalling devices giving orders and authorising the production or movement of an item.

3 RESULTS AND DISCUSSIONS

3.1 Presentation

This section will show the results of the research into the company Supermercado x (fictitious name), describing how stock management is carried out with the aid of applied automation and the results in line with the objectives of this work with regard to planning needs and stock control.

From 10 to 15 February 2016, interviews were carried out using questionnaires (Appendix A), with closed-ended and open-ended questions for the following departments: butchery, billing, cold cuts, produce and grocery, which are responsible for controlling stock and purchasing products in inventory management. The heads of the company's Information Technology (IT) and Human Resources (HR) departments were also interviewed. They all use the software implemented for stock control in the company.

3.2 Company characterisation

The company Supermercado x (fictitious name), which operates in the retail trade in general, with a predominance in food products, is located in the municipality of Cacoal/RO. It was founded on 5 October 1962 in the small town of Aparecida d'Oeste, São Paulo, working in the dry and wet goods sector, and later also exploring cereals and coffee, where trade was booming at the time. The company grew and needed to expand.

With a great entrepreneurial vision, in January 1986 the directors set off in search of a new path. They visited other states and got to know several cities, until they arrived in Cacoal-RO, where they settled, finding a hospitable population, a municipality with fertile land and booming business. They had many difficulties at first, but they never gave up and kept their sights set on conquering this promising new market.

According to the Human Resources management department at Supermercado x , the company is considered one of the best in the region, with a sales area of approximately 1650 metres^2 , with twenty-one (21) checkouts, with a mix of twenty-two thousand (22,000) registered items, unique innovation in the state with a rotisserie with a variety of foods, vegetables and seafood at affordable prices.000) registered items, unique innovation in the state with a rotisserie with a variety of foods, a complete fishmonger with seafood at affordable prices, a market with the best options of fruit, vegetables and greens, a butcher's shop that caters for its customers with special cuts of meat and much more, generating more than 190 direct jobs, benefiting the municipality.

The company is always looking to invest, training its employees, trying to satisfy its customers, with a greater mix, better service and comfort, always seeking to innovate and has expanded and modernised its shop.

The company's mission is to offer quality products and services, with variety and differentiated service, conveying credibility to our customers. Its vision is to be a benchmark supermarket, becoming the most admired retail company in the state of Rondônia. Its values are: valuing the customer, respecting and encouraging the development of our employees, teamwork, harmony and co-operation, transparency and honesty, humility and gratitude, focus on the quality of products and services, productivity, social responsibility and commitment to the environment

3.3 Description of results

Every day, around twelve lorries unload goods at two docks, which are first sent to the sorting and checking department, after which the billing department enters the products into the system, each product is then sent to stock and then to the supermarket shelves for sale. The company only has one stock, in which each product is sold according to its order of entry, first in, first out, taking into account expiry dates. The company's stock policy is based on the resupply interval for each product, because each department has its own internal purchasing control. All stock is managed with the help of software that has a database with all the items registered.

The implementation of stock management software has taken place since the company opened in the state of Rondônia, but in a more simplified way. Before, there was no idea of what the company had in terms of assets, and today it is possible to have the real value of stock, credit, debit, accounting, tax and control of the entire company structure.

The programme implemented works by means of Kanban, i.e. when stock is low, the system signals red, showing the need for a new order through the department responsible for the product.

When the system indicates a yellow colour, the stock level is medium, requiring attention, and the green colour indicates stable stock, within the quantities desired at the time.

The main purchases were high-performance computers for storing large amounts of information and for faster data processing for the user.

The difficulties faced with implementing the software were the cost of suitable machines for support and the required databases, and the inclusion of all the company's active items in the system. And after implementation, problems arose with the supply interval for each product and difficulties in keeping registration data up to date.

The reasons that led the company to adopt the use of technology for stock management was because buyers could have a more accurate view of what was in stock and how much they had to buy. The results achieved with the implementation of commercial automation in stock management reached 100 per cent, meeting the expected expectations.

Below are the graphs of the company's application of the questionnaire for the departments: butchery, billing, cold cuts, fruit and groceries.

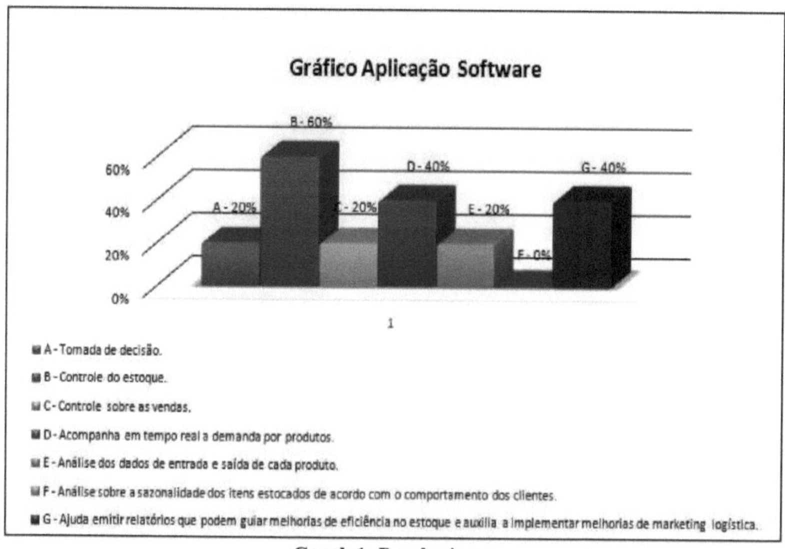

Graph 1: Purchasing management.

SOURCE: PREPARED BY THE AUTHORS, 2016.

Graph 1 shows how automation helps in the development of the company's inventory management activities in the five departments interviewed: 20% in decision-making, 60% in inventory control, 20% in sales control, 40% in real-time monitoring of demand for products, 20% in analysing incoming and outgoing data for each product and 40% in issuing reports, making it possible to drive improvements in inventory efficiency by implementing marketing and logistics improvements, as can be seen in the graph.

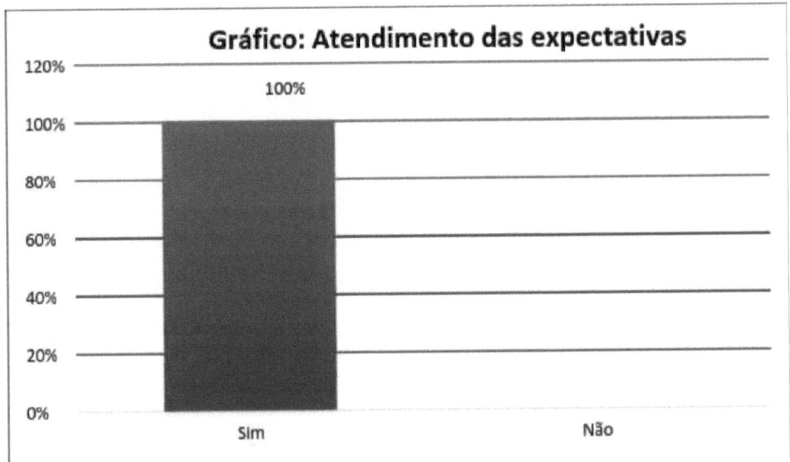

Graph 2: Meeting employee expectations.

Graph 2 shows the percentage of fulfilment of employees' expectations regarding the use of the software in the development of their activities.

3.4 Discussion of results

The survey shows that with the implementation of commercial automation the company was able to obtain 100% of the results, having real control of the value of stock, credit, debit, accounting, tax and the entire structure of the company. The departments interviewed are satisfied with the software used in the company. The main problems and changes they had to make in order to implement the technological stock management system were acquiring high-performance computers to store large amounts of information.

According to the results obtained, commercial automation applied to the stock management of a supermarket provides better control of the quantities of products in stock, facilitating the development of purchasing activities.

4 CONCLUSION

The study presented here met the objectives of identifying the results obtained with the application of automation in stock management at the supermarket; verifying the improvements obtained with the implementation of automation in stock management, what problems are faced after the implementation of automation and analysing the results obtained with the application of automation in stock management at supermarket X.

The results obtained with the implementation of the stock management software gave the company control of its assets, the real value of its stock, through the programme implemented with the help of Kanban, in order to have a more accurate view of what it has in stock and how much it will have to buy, with the acquisition of high-performance computers that generated high costs with machines to support and store databases. Difficulties have arisen in keeping registration data up to date.

In order to have an exact view of what is in stock, the company opted to use technology for stock management. In this way, it achieved 100% stock control, meeting expectations.

According to the research carried out at supermarket X, it concludes that automation has contributed to better stock management so that purchases comply with the company's stock policies.

5 REFERENCES

BASTOS,R. M; SEGRE, L. M. , 1999. **Productive modernisation in supermarkets: adoption of information and communication technology.** Available at: <file:///C:/Users/Acad%C3%AAmico/Downloads/3504072. pdf>. Accessed on: 25/11/2015.

BAZZOTTI, Cristiane ; GARCIA, Elias. 2007. **The importance of the management information system for decision-making.** Available at: <http://www.waltenomartins.com.br/sig_texto02.pdf>. Accessed on: 07/02/2016.

DUARTE, Tiago, 2013. **The automation engineer and control.** <https://www.academia.edu/8476254/O_ENGENHEIRO_DE_CONTROLE_E_AUTO MA%C3%87%C3%83O>. Accessed on: 28/11/15.

FIGUEIREDO, Ricardo Motta, 2010. **Implementation of the pull system in a company producing capital goods.** Available at: <file:///C:/Users/Acad%C3%AAmico/Downloads/Kanban.pdf>. Accessed on: 25/11/15.

FLEURY, P. F; WANKE, P.; FIGUEIREDO, K. F. **Logística Empresarial: uma perspectiva brasileira**. São Paulo: Atlas, 2000.

OHNO, T. **The Toyota Production System: beyond large-scale production**. Porto Alegre: Bookman Companhia, 1997.

PORTER, M.E. **Competitive Strategy: techniques for analysing industry and competition.** Editora Campus Ltda, Rio de Janeiro, 1985.

SANTOS, Angela. M. M. M, 2. **RESTRUCTURING OF THE RETAIL AND SUPERMARKET TRADE.** 2002. Available at: <http://www.bndespar.com.br/SiteBNDES/export/sites/default/bndes_pt/Galerias/Arq uivos/conhecimento/bnset/set903.pdf>. Accessed on: 25/11/15.

SILEVIRA, LEONARDO; LIMA, WELDSON Q. 2003. **A brief conceptual history of Industrial Automation and Networks for Industrial** AutomationAvailable **at** : <http://www.dca.ufrn.br/~affonso/FTP/DCA447/trabalho1/trabalho1_13.pdf>. Accessed on: 24/11/15.

SUCUPIRA, Cezar. 2009. **How to draw up stock management policies.** Available at: <http://ogerente.com.br/img_artigos/logistica/artigo- logistica-politicas-de-estoque.pdf>. Accessed on 29/11/15.

TUBINO, D. F. **Sistema de produção: a produtividade no chão de fábrica.** Porto Alegre: Bookman Companhia, 1999.

VILELA, Rodrigo S. 2006. **Strategic tools and technologies used in commercial automation for efficiency in supermarkets.** Available at : <http://www.em.ufop.br/cecau/monografias/2005/RODRIGO%20S.%20VILELA.pdf>. Accessed on 24/11/15.

WANKE, Peter. **Inventory Management in the Supply Chain: Decisions and quantitative models.** 3rd edition. Publisher: Atlas - São Paulo, 2011.

CHAPTER 8

ANALYSING THE BARRIERS ENCOUNTERED BY AN AUTOMATION PROJECT COMPANY IN RONDÔNIA

Carlaile Largura do Vale
Karoline Quiteria M. Borba
Marly Agatha E. Galheno
Nícolas Alessandro de Souza Belete
Tatiane Aparecida de Lazari

SUMMARY

Automation is increasingly being used by various companies and industries, as well as in homes through the concept of home automation. This wide range of services and products offered by automation systems makes their use essential for safety, comfort, ease, speed and confidence in carrying out various tasks and activities in people's routines. On the other hand, the culture of using human labour for almost all tasks is still present in many regions, and adhering to a new perception can be difficult at first. For this reason, many automation companies encounter a number of barriers in their operations, including those related to geographical location, because considering the size of Brazil, it is known that regions that are not part of the major economic centres are relatively distant, making it difficult to access certain services and products resulting from technological innovation. This is why this study sought to analyse the barriers encountered by an automation project company in the state of Rondônia and how it deals with these situations. To this end, a questionnaire was administered to an employee responsible for the organisation used as the object of study. The results showed that the main barriers are based on the general public's lack of information about the implementation of industrial and residential automation, including factors such as affordability, ease of use, among others. To get round these situations, the company reports on the need to expand knowledge and information about the benefits and importance of automation, as well as how it works.

Keywords: Automation. Barriers. Rondônia. Technological innovation.

1 INTRODUCTION

With the development of new techniques for improvement and the search for greater productivity in various areas of production, and even in the services market, it is necessary to re-evaluate the problems encountered by companies seeking to stand out in the market through automation, thus having an edge over their competitors, given that the practice of using certain strategies is scarce in certain areas.

For each area to be studied, there is a different applicability of automation, i.e. specific techniques and methods that can be decided on by means of the quantity to be produced, the type of item, or similar production processes, thus resulting in work optimisation and high productivity as a result of pre-established planning that is carried out on the basis of the specific characteristics of each branch (TUBINO, 2006).

It was therefore necessary to find out about the barriers faced by a company in the building industry, taking into account the northern region, more precisely in the state of Rondônia, a company in Cacoal, which provided the information by means of a questionnaire.

There are a number of obstacles that companies developing automation projects face in any region of the world, whether it's the difficulty of accessing technological innovations, the low levels of investment available, or even the culture in which the society in question is accustomed. Within this context, what are the biggest barriers encountered by a company developing automation projects for corporate and residential buildings in the state of Rondônia? To answer this question, this study aims to analyse the biggest barriers encountered by a company that develops automation projects in corporate and residential buildings in the state of Rondônia. From this point of view, it is of great relevance to present the main reasons that slow down the application of automation in the workplace, and what possible changes could make it possible to reduce these obstacles.

2 THEORETICAL FRAMEWORK

The topics below present concepts and information on subjects that are important for understanding this study, which includes the types of production process and the entire relationship between automation in corporate and residential buildings, as well as project companies and the challenges they face in inserting automation into today's culture.

2.1 Production process models

Each process model acts in a different way in the organisation of activities and operations, according to their different aspects of volume and variety, since the quantity of these items has a direct impact on the organisation of production, and on planning, in order to establish each stage of the procedures (SLACK et al, 2002).

According to Tubino (2007), production systems can be classified in different ways, each following the nature and level of standardisation of the products, or by the types of processes involved during the production stages. Tubino (1999) states that standardised products indicate a higher degree of equality, while made-to-measure products are created specifically for each consumer.

2.1.1 Project processes

Project-type processes manufacture particular, unique products that are customised and take a long time to make. Due to their unique characteristics, the products have low volume and a wide variety in production. An example of this type of project is the production of ships and aeroplanes, where detailed planning takes place so that each stage of the process is directed and executed in a special way for each product (SLACK et al, 2002).

2.1.2 Jobbing processes

Jobbing processes are characterised by a high variety of products and a low volume of production. They differ from process design in that in process design each stage of production is geared exactly to that type of product, while in jobbing there is a sharing of resources in the stages with other products. Examples of jobbing include tailors and furniture restorers (SLACK et al, 2002).

2.1.3 Batch or batch processes

Batch processes have a smaller variety than Jobbing processes, and a greater volume of production, because their products are manufactured in batches, as the name implies. Examples include the manufacture of shoes and clothes (SLACK et al, 2002).

2.1.4 Mass production processes

Mass production processes are characterised by the fact that they produce a large volume of products in a linear, non-flexible sequence, with little variety, i.e. the production steps are repetitive and predictable, thus producing several of the same products. An example of this is a car or electronics factory (SLACK et al, 2002).

According to Corrêa & Corrêa (2004), mass processes are defined by the large-scale production of discrete, highly standardised products, flowing synchronously from workstation to workstation at a pre-established rate.

2.1.5 Continuous processes

Continuous processes have a lower variety than mass production processes and a higher volume, because their production operates in a continuous stream, without stops, highly predictable according to the characteristics of the production. Examples include petrochemical refineries and electricity supply (SLACK et al, 2002).

2.2 Automation

Ribeiro (2001) states that the concept of automation is based on the use of machines to replace human or animal labour, in an automatic system or via remote control where there is little or no direct interference from human beings. In this way, the system reacts by acting on its own, based on a certain programme of activity at a specific time or reacting to certain conditions.

Automation has been developing at a rapid pace, mainly due to the repercussions of its benefits, especially in the industrial sector, where any competitive advantage is sought. This is because the applications are not only based on replacing human activities that are considered repetitive, exhausting and dangerous, but also because automation optimises spaces and processes, reducing time, excess costs and losses due to human error (PAREDE and GOMES, 2011).

In this context, Parede and Gomes (2011) point out that in the industrial sector the most prominent piece of equipment is the programmable logic controller (PLC), which has revolutionised industrial controllers by replacing large electromechanical relay cabinets with extensive wiring with discrete components with a low integration scale. From this, it can be seen that automation began as a solution for industrial improvement, where initially it was only viable for organisations, a fact that, with the evolution of automation itself, proved to be incoherent. For a better understanding of the evolution of automation, table 1 shows the main events in the technological evolution of the PLC, the main piece of automation equipment.

Table 1 - Technological evolution of the PLC

Decade	Events
1960	The emergence of the PLC to replace control panels with electromechanical relays, energy savings, ease of maintenance, reduced space and lower costs.
1970	0 PLC acquired timing instructions, arithmetic operations, data movement, matrix operations, programming terminals, PI D analogue control. At the end of the decade, communication resources were incorporated, enabling integration between distant controllers and the creation of various proprietary communication protocols (incompatible with each other).
1980	Reduction in physical size due to the evolution of electronics and the adoption of intelligent I/O modules, providing high speed and precise control in positioning applications. Introduction of software programming in microcomputers and the first attempt to standardise the communication protocol.
1990	Standardisation of programming languages under the IEC 61131-3 standard, introduction of human-machine interfaces (1 HM), supervisory and management software, fieldbus interfaces and function blocks.
Today	A concern to standardise communication protocols for PLCs so that they are interoperable, enabling one manufacturer's equipment to communicate with another's, which facilitates automation, management and the development of more flexible and standardised industrial plants.

As can be seen in table 1, automation began with the PLC, but with advances in technology not only has this equipment undergone transformations aimed at improvement, but also, according to Parede and Gomes (2011), there has been the introduction of new products and PLC communication networks so that activities carried out by processors that were located at the intelligent input and output are now performed in software by the central processing unit.

Within industrial automation, there are classifications of types and means defined by basic characteristics, components and indications for use, which according to Parede and Gomes (2011), are as follows:

- a. Pneumatic controls (directional control valves);
- b. Electrical and electro-pneumatic controls with PLC;
- c. Use of PLC or computer;
- d. Digital distributed control system (DCS);
- e. Industrial environment (IP or NEMA).

Observing that there are various types and options of automation, one must then understand the process in which one wishes to implement it in order to analyse the cost-benefit of each technology presented, always focusing on the simplicity, practicality and feasibility of each one, as these are determining factors for the success and efficiency of the project. Good sense and planning are also essential during this implementation (PAREDE and GOMES, 2011).

2.2.2 Residential buildings

Automation in residential buildings consists of integrated technological systems that perform functions and commands based on programmable instructions and provide services to satisfy needs related to security, communication, comfort and energy management in a residential environment (MURATORI, 2010).

According to Muratori (2010), the equipment and systems generally used for home automation can be divided into four classifications:

Electrical installation automation: comprising control and supervision software, relays, interfaces and timers, various sensors, gas, smoke and flood detectors, among others.

Universal remote controls: includes fixed (wall-mounted) touchscreen controls, mobile panels, interfaces needed for integration, home automation systems, integration software licences (iphones, mobile phones, pocket PCs, etc).

Energy management: Includes energy consumption managers, metering of water and gas consumption, electrical protection equipment and alternative energy generation.

Accessories and complements: this class is based on various automation items that don't fit into the previous classifications, such as motorisation of blinds, awnings and curtains, heated floors, central vacuum aspiration, mirror defoggers, automated irrigation, electric locks, access control equipment (biometric reading, keypads), media centres, network assets (switches, routers), telephony and intercom (conventional and IP), among others.

2.2.3 Project companies

Automation project companies in Brazil are gradually acquiring characteristics similar to those of companies in more developed markets, so that the channel is positioned as shown in figure 1.

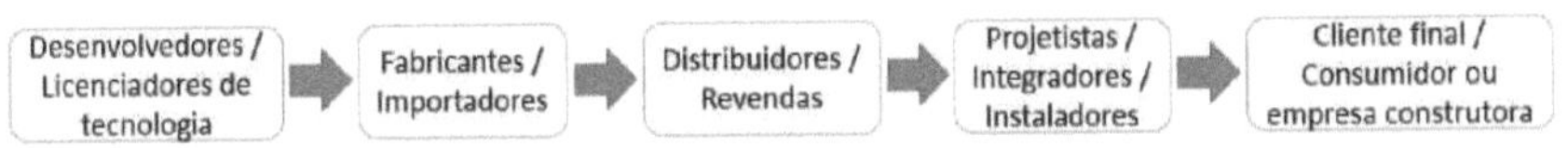

Figure 1. Flowchart of the project companies market.

SOURCE: MURATORI, 2010.

2.2.4 Automation challenges

There are many challenges that automation faces if it is to be accepted and implemented in both the industrial and residential sectors. This, according to Neves et al (2007), is the result of factors involving the social sphere,

society's education in the face of technological evolution, safety, reliability, among others. In this context, the author also mentions that the ten main automation challenges are defined as:

Technical Training of Professionals and Education of Society in the Technological Evolution Provided by Automation;

Safety and Reliability in Critical Systems;

Optimisation of Information to Provide an Appropriate Human Machine Interface;

1. Pattern Recognition;
2. Fault identification in automation systems;
3. Secure communication between heterogeneous devices;
4. Home Automation Systems;
5. Real-Time Information Management;
6. Applications in Medicine;
7. Social and Environmental Impacts of Automation.

3 METHODOLOGY

According to Michel (2005), research is the search for the study, analysis, interpretation and description of knowledge on a particular subject that has been generated by inexhaustible approaches. This research is of a basic, descriptive nature, with a qualitative and quantitative approach and field research. The first stage consisted of carrying out a bibliographical survey of the topics covered. Based on the information obtained, a questionnaire was drawn up and applied, shown in Appendix A, consisting of open questions to the organisational head of an automation project company in the state of Rondônia. The final stage was based on compiling, analysing and discussing the results obtained from the previous stages. Considering the ethical aspects, the integrity and confidentiality of the information was respected, as well as the institutions and subjects involved in the research based on the principles accepted as correct and moral.

4 RESULTS AND DISCUSSIONS

The questionnaire revealed that there is little information about home and industrial automation in the Rondônia region, which results in the population having little knowledge of this service, which also includes various products. On the other hand, this deficiency has been reduced considerably, mainly due to the expansion of knowledge via the internet and the dissemination of the existence of such services through cinematographic products (films, series, programmes, etc.). As a result, there has been progress, albeit small compared to the rest of the country and the world, and currently the services most sought after by consumers in the company are automation solutions, CCTV (Closed Circuit Television), CATV (Cable TV), fire detection, access control, structured cabling, audio and video. For such installations, there is a need for the client to communicate with the company, in which they present their wishes in relation to the service/product, so that the company can analyse how the appropriate automation will be structured with the other necessary information in relation to values, maintenance, accessibility, equipment options and services, among others. Generally, the company's clients are organisations that provide health, wellbeing, safety and ease of use between processes and communication, as well as people who have a high level of technological knowledge and/or interest. As such, there are generally no major difficulties for clients in adapting to such systems, with a few exceptions. These exceptions are usually people with a low level of technology and/or education, or of an older age, as this audience presents a greater barrier to adherence. For this reason, in some situations it is difficult to identify faults with the systems implemented, mainly due to incorrect use or not understanding how they work.

The biggest difficulties in implementing home and industrial automation are related to a lack of knowledge about its existence, importance and benefits. There is also the fear that such systems are highly expensive and difficult to understand and handle, which for those looking for information, these facts prove to be fallacious. Many people also believe that automation only provides services and products that are unnecessary and for entertainment purposes or luxury items, also wrongly, as they are mainly used for security purposes. For this reason, the activities carried out to overcome these barriers are mainly the dissemination of information and knowledge about automation, its benefits, among others, through the internet, word-of-mouth marketing, consultancies, etc. There are certain social impacts observed in the implementation of automation in homes and industries in relation to people's behaviour, the culture of activity procedures and the very breaking of previously existing paradigms.

As far as the state of Rondônia is concerned, there is a lot of progress to be made when it comes to technology and automation, so the state is not compatible with the other regions of the country and, consequently, with the rest of the world. The distance between the country's major economic centres is one of the factors interfering with the spread and use of automation in the state. This factor can be explained by the encouragement of both the government and organisations to adopt automation, which is much more prevalent in the south and southeast. The impact of this distance lies in the price of the technology, which is comparatively higher due to the lack of competition and people's contact with automation. The low number of product and service options is also due to the distance between the state and the country's major centres. In order to reduce the impact of this factor, the interviewee mentioned that it is fundamentally important to constantly keep up to date with innovations and to continually seek out information and knowledge on the subject, while also focusing on disseminating this to the public in order to bring integration to the region in relation to the subject and levelling the playing field in relation to the country.

In order to increase interest in implementing automation, the main element is to show the benefits that are gained in all aspects, such as security, financial, ease, optimisation of tasks and time, comfort, entertainment, among others, i.e. pointing out that the cost-benefit ratio is positive. The fact that this is the near future of society and that automation will be increasingly present, becoming fundamental to a company's competitiveness, is also influencing the increase in interest. As for the decline in interest, this is generally caused by bad experiences of consumers with companies that have offered some form of automation implementation, but which do not value the quality and functionality of their services and products, as well as after-sales, maintenance, training (when necessary) in relation to use, among others, which also includes a lack of information.

5 CONCLUSION

Based on all the information obtained in the questionnaire, it can be concluded that the biggest barriers encountered by the company that implements home and industrial automation services is the lack of knowledge and information that people in the region have about this service. This is very common in many regions of the country that do not belong to or are far from the country's economic and industrial centres, which are concentrated in the south and south-east. As a result, there is a strong culture that automation is not only something that can only be imagined, but is also an extremely expensive and inaccessible service. Even though the internet provides many examples and information about automation, it is known that breaking cultures is often complicated and time-consuming, making the process of embracing this new reality slow.

In order to reverse this situation, it is important to have strong marketing strategies in order to consolidate the concept of automation in everyday life, as well as to constantly keep up to date with existing innovations and seek to expand the information and knowledge that has changed people's view of the services and products on offer. The growth of the economy and the interest in expanding business in the region, as well as the perception of market insertion as being advantageous for large companies, implies the need to adopt services that provide competitiveness, one of which is automation. As a result, the use of these services and products tends to increase, as does the dissemination of information about their importance, benefits and information in general. Therefore, such barriers tend to be reduced, even eliminated, with the economic advancement of the state itself.

REFERENCES

CORREA, H; CORRÊA, Carlos. **Production and operations management**: manufacturing and services: a strategic approach. São Paulo: Atlas, 2004.

MICHEL, M. H. **Metodologia e pesquisa científica em ciências sociais.** São Paulo: Atlas, 2005.

MURATORI, J. R. **Residential Automation:** history, definition and concepts. Santa Cecília, SP: Portal O Setor Elétrico, 2010.

NEVES, C., DUARTE, L., VIANA, N., FERREIRA, V. **The ten biggest challenges in industrial automation:** the outlook for the future. II Research and Innovation Congress of the North-East Technological Education Network, João Pessoa, PB, 2007.

PAREDE, I. M., GOMES, L. E. L. **Eletrônica:** Automação Industrial. São Paulo: Padre Anchieta Foundation, 2011. Interactive Technical Collection. Electronics series, v. 6.

RIBEIRO, M. A. **Industrial Automation.** Salvador: Tek Treinamento & Consultoria Ltda, 2001. 4 ed.

SLACK, N., CHAMBERS, S., JOHNSTON, R. **Production management.**

Translation by Maria Teresa Corrêa de Oliveira. - 2. ed. - São Paulo: Atlas, 2002.

TUBINO, D. **Sistemas de Produção**: A produtividade no chão de fábrica, Porto Alegre: Bookman, 1999.

TUBINO, D. F. **Manual de planejamento e controle da produção** - 2. ed. - São Paulo: Atlas, 2006.

TUBINO, D. F. **Planejamento e controle da produção:** teoria e prática. São Paulo: Atlas, 2007.

yes
I want morebooks!

Buy your books fast and straightforward online - at one of world's fastest growing online book stores! Environmentally sound due to Print-on-Demand technologies.

Buy your books online at
www.morebooks.shop

Kaufen Sie Ihre Bücher schnell und unkompliziert online – auf einer der am schnellsten wachsenden Buchhandelsplattformen weltweit! Dank Print-On-Demand umwelt- und ressourcenschonend produziert.

Bücher schneller online kaufen
www.morebooks.shop